Harriet Randolph

Laboratory directions in general biology

Harriet Randolph
Laboratory directions in general biology
ISBN/EAN: 9783337215262
Printed in Europe, USA, Canada, Australia, Japan
Cover: Foto ©berggeist007 / pixelio.de

More available books at **www.hansebooks.com**

LABORATORY DIRECTIONS

IN

GENERAL BIOLOGY

BY

HARRIET RANDOLPH, Ph.D.
*Demonstrator in Biology and
Reader in Botany, Bryn Mawr College*

"Willst Du in's Unendliche schreiten,
Geh' erst im Endlichen nach allen Seiten."
—GOETHE.

NEW YORK
HENRY HOLT AND COMPANY
1898

PREFACE.

THE following laboratory-directions have been prepared for a course in General Biology that extends throughout the collegiate year, and includes work in the laboratory for about six hours each week.

Experience has shown that there are certain practical advantages in beginning with the fern and the earthworm, and that the results are not at all inferior to those that follow from the more logical method of beginning with the simplest organisms.

The plan of this course is similar to that suggested by the "General Biology" of Sedgwick and Wilson; and as the directions have been used with large classes for several years, it is believed that they are entirely practicable.

In determining the amount of work to be done upon each form, an endeavor has been made to include the most important points and at the same time to keep each in its proper relation to the other parts of the course as a whole. But any suggestion of improvements that may occur to any one using the directions would be gladly received by the author.

The following list of hours has been found by experience to represent approximately the time needed for the different subjects:

	Hours.		Hours.
Preliminary practice with microscope	3	Lobster	4
		Moss	5
Fern	10	Stem, bud, leaf	4
Earthworm	10	Flower	4
Amœba	2	Seeds	2½
White blood-corpuscles	1	Seedlings	4
Hæmatococcus	4	Seed-contents	4
Paramœcium	2	Circulation, etc., of protoplasm	2
Vorticella	2		
Yeast	5	Karyokinesis	2
Penicillium		Frog	12½
Lichen	4	Fish	2
Mushroom		Pigeon	6½
Bacteria	2	Rabbit	10½
Spirogyra	2	Embryology:	
Hydra	4	Frog	6½
Mussel	5	Chick	16

At the present time laboratory-directions in a course in General Biology necessarily presuppose the use of subjects and to some extent of methods employed by previous writers. For the mode of presentation of certain topics my acknowledgments are due especially to Prof. E. A. Andrews and to Prof. V. M. Spalding.

To Prof. T. H. Morgan, who has been so good as to read and criticise the directions for Invertebrates, Plants, and Embryology, and to Prof. J. W. Warren, who has kindly advised me in regard to the Vertebrate work, it gives me great pleasure to acknowledge my indebtedness.

H. R.

BRYN MAWR COLLEGE, June 1897.

TABLE OF CONTENTS.

	PAGE
INTRODUCTION	1
PRELIMINARY PRACTICE WITH THE MICROSCOPE	5
FERN	7
EARTHWORM	13
AMŒBA	24
WHITE BLOOD-CORPUSCLES	26
HÆMATOCOCCUS	27
PARAMŒCIUM	30
VORTICELLA	33
YEAST	36
PENICILLIUM	39
LICHEN	41
MUSHROOM	43
BACTERIA	45
SPIROGYRA	48
HYDRA	52
FRESH-WATER MUSSEL	58
LOBSTER	66
MOSS	75
SPERMAPHYTES	79
Stem	79
Root	82
Bud	82
Leaf	83
Flower	84
Seeds	87
Seed-contents	90
Seedlings	91

	PAGE
ROTATION AND CIRCULATION OF PROTOPLASM	93
KARYOKINESIS	94
FROG	95
FISH	113
PIGEON	116
RABBIT	128
EMBRYOLOGY	143
Frog	142
Chick	146

GENERAL INTRODUCTION.

1. THE MICROSCOPE AND ITS USE.

The microscope * consists of a *stand*, a *tube*, and *lenses*. Attached to the column of the stand is a horizontal plate, the *stage*, which holds the object to be examined.

In the middle of the stage is a round hole through which light can be directed upward by a *mirror* which is attached to the stand below the stage. The amount of light can be regulated by *diaphragms*.

The stand gives support above the stage to a vertical cylinder in which the *tube* slides up and down.

The combination of lenses to be inserted into the upper end of the tube is the *eyepiece* (or ocular); that at the lower end the *objective*.

The objective AA is a low power, D is a high power.

Objective AA, eyepiece 2 magnifies about 50 diameters.

Objective AA, eyepiece 4 magnifies about 90 diameters.

---

* The description applies to the Zeiss microscope, Stativ VII, with oculars 2 and 4, objectives AA and D.

Objective D, eyepiece 2 magnifies about 240 diameters.

Objective D, eyepiece 4 magnifies about 420 diameters.

To be seen clearly an object must be at a definite distance from the objective. Moving the lenses to the proper place is called *focussing*.

For finding the focus there are two adjustments. The *coarse adjustment* is used especially with the low power; it may be used also with the high power for finding the approximate focus; it consists in sliding the tube up or down with the fingers. This should always be done with a slight twisting movement. The *fine adjustment* is effected by means of a screw with a milled head behind the tube. It is used to focus the lens exactly after the object has been brought into view with the coarse adjustment.

The following rules should be carefully observed:

1. Always examine an object first with the low power (AA).
2. Never use the high power (D) unless the object is covered with a cover-glass. Be very careful never to let the objective touch the cover-glass.
3. Remember that objects can be brought clearly into view only by the hand, not by the eye. If the object is not dis-

tinctly visible do not strain the eye to see it, but by means of the fine adjustment alter the focus of the microscope. In order to bring different parts of an object into view the focus must be continually changed, and for this reason the hand should always be kept on the fine-adjustment screw. If this rule be observed there is little danger of injuring the eyes.

If the object cannot be brought clearly into view by focussing, see whether the mirror reflects as much light as possible, that the objective is not wet, and that the cover-glass is clean.

4. The lenses must be perfectly clean. Dust on the lens obscures the image, and may strain the eye. To clean a lens, breathe upon the glass in order to dampen it, and then wipe it with a soft, fine cloth kept for this purpose alone. A lens should never be touched with the fingers, as these often leave ineffaceable marks.

5. Look through the microscope sometimes with one eye and sometimes with the other, and keep both eyes open; the strain upon the eyes is by this means much lessened.

6. Lenses, when not in actual use, should be protected from dust and injury.

7. The tube of the microscope should always slide smoothly. If it does not, take it out, wipe it clean, and apply a little vaseline.
8. When lifting the microscope, do not take hold of the column above the stage, as this is apt to disarrange the fine adjustment.
9. When not in use, the microscope should be put into its box (observe paragraph 8). Each objective should be put into its own brass box with the lid that corresponds (otherwise the thread of the screw may be injured, and the boxes cannot be tightly closed, to keep out moisture, etc.).

II. On Drawing.

Everything of importance should be recorded by sketches or diagrams and by notes. Try to show in the sketches as clearly and accurately as possible the essential features of what has been seen, and add concise explanations. In order to show the natural relations of the parts it may be necessary to combine several views of the object.

Draw on a large scale; the sketches cannot be too large.

It is best to draw to scale, i.e., either the exact size or half or double, etc., the size of the object.

Name the parts represented in the drawing, and

mark the scale adopted, as, e.g., × 2 when the drawing represents the object twice its size.

The outlines should first be traced very lightly with a soft pencil, so that they may be readily altered. When the sketch is complete the outlines may be gone over with a harder pencil or with ink, and the different parts may be made more distinct by slightly shading them with colored crayons.

Outlines should not be sketched in color. Certain colors should be used for special organs or tissues.

Preliminary Practice with the Microscope.

I. Scales from the wing of a butterfly.

Take with a scalpel some of the colored particles from the wing, place them upon a slide and examine with AA 2. Compare the object as it appears when illuminated by the the mirror ("transmitted light"), and when the light from below is cut off ("reflected light"). Examine successively with AA 4, D 2, D 4, and make a series of outline sketches of a few of the scales to show the difference in magnifying power.

II. Put a few fibres of wool into a drop of water upon a slide. Put on a cover-glass and examine first with a low and then with a high power. Note the form of the fibre by carefully focussing at different levels. Sketch (D 2).

III. Examine bubbles of air in water. To obtain these, support a cover-glass by a thick piece of filter-paper under one edge, run a drop of water under the cover glass, and then tap the glass.

The cover-glass must be supported so that the bubbles may be spherical. Sketch (AA 2). (Keep for comparison with IV.)

IV. Examine a small drop of oil suspended in water with the cover-glass supported. Compare the air and the oil, and note the differences in the appearance of each upon focussing up and down. Sketch.

V. To make a scale. Find on a stage micrometer,* with the low power, the lines ruled one millimetre and one tenth of a millimetre apart. Draw some of these as they appear projected upon paper at the base of the stand. While doing this keep both eyes open and the head motionless. Draw several sets of lines, and then transfer the most accurate set to a piece of cardboard.

Make a scale for the combinations AA 2, D 2, and D 4; label each with the names of lenses used, and the actual size of the micrometer spaces.

---

* Zeiss's Object-micrometer, No. 26, b.

FERN.
PTERIS AQUILINA.

A. ASEXUAL GENERATION.
I. External characters.
1. **Rhizome**—the brown, underground stem with a ridge on each side.
2. Roots ; springing from the rhizome.
3. The leaves or **fronds**: arising from the rhizome at intervals along the lateral ridges.
   (a) The subdivision of the frond : the leafstalk or **stipe**, and the leaf proper or **lamina**. The lamina is subdivided like a feather (pinnately) into **pinnæ**; each pinna is pinnately subdivided into **pinnulæ**, the pinnulæ in some cases into **lobes**, and these into **lobules**.
   (b) The **sorus**: a band of **sporangia** under the edge of some of the pinnules.
   (c) The **indusium**: the membrane covering the sorus.
4. The nodes and internodes of the rhizome.
5. The growing point at the end of the rhizome.
Sketch, showing these parts.

II. The rhizome.
1. *Cut one end smooth with a razor and sketch the cut surface as seen with the naked eye.*
   (a) The outer brownish layer : **epidermis** and **subepidermis (sclerotic parenchyma)**; the

subepidermis thinner at the lateral ridges.
(b) The light-colored ground-substance or **fundamental parenchyma**, forming most of the stem.
(c) The dark interrupted ring of **sclerenchyma (sclerotic prosenchyma)**.
(d) Small patches of sclerenchyma outside the main ring.
(e) Yellowish tissue (**fibrovascular bundles**) lying inside and outside the ring of sclerenchyma.

2. *Cut a small piece of rhizome in a longitudinal vertical plane.*

Find on the cut surface (a), (b), (c), (d), (e), in II. 1.

Sketch.

3. *Cut a thin transverse section of rhizome, mount in water, and examine with the low power.*

Find the different kinds of tissues, as in II. 1, (a), (b), (c), (d), (e).

Make a diagram to show the outlines of the different regions.

4. *Treat the section with iodine and observe the result. To treat with iodine put a drop or two of iodine on the slide close to the cover-glass, and on the opposite side of the cover-glass draw away the water from under the cover-glass with a small piece of filter-paper.*

5. Examine a prepared stained transverse section of rhizome. Study with the high power.
   (a) Find again the tissues in II. 1, (a), (b), (c).
   "(a)" The epidermis: a single layer of cells with very thick walls; the subepidermis (sclerotic parenchyma): with thick-walled cells.
   "(b)" The fundamental parenchyma: large thin-walled cells with protoplasm, nucleus, and starch-grains.
   "(c)" Sclerotic prosenchyma: smaller cells with thick walls.
Sketch a few cells of each kind.
   (b) In the fibrovascular bundle note:
      1. On the outside a single layer of flattened cells with brownish walls and apparently empty (**bundle-sheath**).
      2. Within the bundle-sheath a layer of parenchymatous cells containing starch (**phloëm-sheath**).
      3. Within the phloëm-sheath is the phloëm, consisting of:
         (a) Small cells with thick walls (**bast-fibres**).
         (b) A single row of large thin-walled cells (**sieve-tubes** or **bast-vessels**).
         (c) Small thin-walled cells lying be-

tween (*a*) and (*b*), and containing starch (**phloëm-parenchyma**).

4. Within the phloëm is the **xylem** or wood. N te :
   (*a*) Large err ty vessels with thick walls (**tracheids**).
   (*b*) Small vessels (**tracheæ**).
   (*c*) Small parenc -cells containing starch, i spaces between the angles of vessels (**wood-parenchyma**).

Sketch the outline of a fibrovasc bundle and some cells of each kind in their relation to one another.

6. In a prepared longitudinal sec. find the kinds of cells just seen in tra section.

Sketch.

7. *Crush a piece of rhizome that has been n cerated, and with the high power find the var kinds of cells now more or less isolated. Sketch.*

III. The Leaf.

1. *Strip off a piece of the epidermis from th ower side of the leaf, mount in water, and no.*
   (*a*) The ordinary epidermis-cells with v outlines.
   (*b*) The **guard-cells**, two surrounding each **stoma**.

Sketch accurately (much enlarged) two guard-cells,

with a few of the ordinary epidermis-cells surrounding.
2. In a prepared vertical section of a leaf note :
   (*a*) The epidermis-cells of upper and lower surface. Look for stomata and guard-cells.
   (*b*) Mesophyll-cells containing much chlorophyll. These cells are arranged near the upper surface of the leaf in compact layers, but near the lower surface they are irregularly placed, with air-spaces between them.
   (*c*) Transverse or oblique sections of the fibrovascular bundles.
Sketch a part showing (a) and (b).
IV. Asexual Reproductive Organs.
   1. *Examine a part of the sorus with a low power without a cover-glass.*
      It is composed of a number of oval bodies: the sporangia, covered by the indusium. *Sketch.*
   2. *Remove some of the sporangia, mount in water, examine with high power, and sketch.* Note:
      (*a*) Form: biconvex bodies, borne on short stalks.
      (*b*) Structure: flat cells, with one row thickened to form a prominent ring (**annulus**) around part of the **capsule.**

(c) Mode of dehiscence (look for one that has opened).
3. Spores. Examine with high power. Note:
    (a) Size. (*Measure.*)
    (b) Form.
B. SEXUAL GENERATION. Prothallium.
 I. Examine a prothallium with the naked eye or a hand-lens, then mount in water, and examine with the lowest power.
    Sketch. Turn it over and sketch again.
II. Reproductive organs.
    1. Examine the lower surface of the prothallium with the high power. Note:
        (a) **Archegonium**: a chimney-like structure, brownish in late stages. In the cavity at the bottom lies the oösphere. *Sketch.*
        (b) **Antheridium**: a rounded projection with an outer layer of cells containing a few chlorophyll-grains. In different stages of development there can be seen in the centre either a single cell or a number of smaller cells (mother-cells of spermatozoids) produced by the division of the central cell. The mother-cells in ripe antheridia contain spirally coiled bodies (spermatozoids). *Sketch.*

EARTHWORM.
LUMBRICUS TERRESTRIS.

A. OBSERVATIONS ON THE LIVING WORM.

I. *Put a living worm upon a sheet of paper on the laboratory table and observe its shape and its movements. (Be careful not to let the worm become dry.)*

   1. Note the changes of form of the anterior and posterior parts of the body as the worm moves over the paper. In what way is this change of shape brought about?

   2. With a blunt instrument touch one end and then the other end of the worm. Which end is more sensitive?

   3. Note the pulsations in the dorsal blood-vessel (visible through the skin along the median dorsal line). What is the direction of the wave of contraction?

B. EXTERNAL CHARACTERS OF THE PRESERVED SPECIMEN.*

I. *With the worm extended in a dissecting-pan and covered with water or with weak alcohol, observe:*

   1. The division of the body by constrictions into a number of segments or **somites** (me-

---

* For method of preservation, etc., see "General Biology," Sedgwick and Wilson, second edition, 1895, pp. 210–213.

tamerism or serial symmetry). Count the number of somites.
2. Modifications of the somites:
    (a) At the extreme anterior end a smoothly-rounded knob, the **prostomium**.
    (b) Swellings on the ventral side of some of the somites between the 7th and the 19th, the **capsulogenous glands**.
    (c) Between the 28th and the 37th somites a swollen region, the **clitellum**.
3. The anterior and posterior ends are not alike (**antero-posterior differentiation**).
4. The dorsal and ventral sides are unlike (**dorso-ventral differentiation**).
5. When the worm is placed with the ventral side down a longitudinal vertical plane will divide it into similar right and left halves (**bilateral symmetry**).
6. The stiff bristle-like **setæ**: felt by drawing the worm gently through the fingers. With the naked eye or a hand-lens note the arrangement of the setæ.
7. The openings to the outside:
    (a) The **mouth**: close to the anterior end on the ventral side below the prostomium.
    (b) The openings of the **seminal receptacles**: on each side between the 9th and 10th and the 10th and 11th somites (four openings; difficult to make out).

(c) The openings of the **oviducts**: one on each side on the ventral aspect of the 14th somite.

(d) The openings of the **vasa deferentia**: a conspicuous slit surrounded by swollen lips, on each side on the ventral aspect of the 15th somite.

(e) **Dorsal pores**: one in each somite after the first two or three, in the median dorsal line.

(f) **Nephridial** openings: one pair in each somite. (These cannot be seen with the naked eye.)

(g) The **anus**: a vertical slit in the posterior (the **anal**) somite.

Make two enlarged drawings (× 3 or × 4) of the anterior end, a dorsal and a ventral view, showing the first forty somites; and a sketch of the posterior end, dorsal aspect, showing the last ten somites.

C. ANATOMY.

I. *Extend a freshly killed worm dorsal side uppermost in a dissecting-pan and fasten it to the wax by a pin at each end (the anterior pin through the prostomium only). Cover the worm with 50 per cent. alcohol. With fine scissors cut carefully through the dorsal body-wall a little to one side of the median line. Note:*

1. The **body-wall** consisting of three layers:
   (a) The **cuticle**.
   (b) The **hypodermis**.
   (c) The muscular layer.
2. The space between the body-wall and the alimentary canal, the body-cavity or **cœlom**.
3. The thin muscular partitions (**dissepiments**) which divide the cœlom into a series of compartments.

II. *Cut through the dissepiments on the left side close to the body-wall; spread out the body-wall and pin it to the wax (putting the pins in obliquely). Similarly for the right side. Note:*
1. The **alimentary canal**: a straight tube extending throughout the whole length of the body.
2. The **nephridia**: delicate, whitish bodies attached by one end to the posterior side of a dissepiment, and by the other end to the ventral body-wall; one on each side of the alimentary canal in each somite.
3. The **seminal vesicles**: three pairs of white bodies lying partly at the sides of the alimentary canal and partly above it, between the 9th and the 12th somites.
4. The **dorsal blood-vessel**: a long tube lying upon the dorsal side of the alimentary canal. It gives off:
   (a) The **aortic arches**: five pairs of large

contractile vessels that pass down around the digestive tract in somites 7 to 11.

(b) Smaller circular vessels: some passing into the walls of the intestine, others along the dissepiments into the body-wall.

5. The parts of the alimentary canal:
   (a) The buccal pouch: an eversible sac into which the mouth-opening leads. (It lies in somites 1 and 2, and should not be dissected out until after the nervous system has been studied. See C. II. 7, (f).)
   (b) The **pharynx**: an elongated thick-walled pouch extending from the 2d to the 7th somite. Numerous small muscles connect it to the body-wall.
   (c) The **œsophagus**: a thin-walled tube much smaller in diameter than the pharynx, extending from the 6th to the 15th somite. (More distinctly seen when being taken out; C. II. 6.)
   (d) The **crop**: a thin-walled dilatation of the alimentary canal at about the 16th somite.
   (e) The **gizzard**: a thick-walled muscular sac at about the 17th somite.
   (f) The **stomach-intestine**: a thin-walled tube

extending from the gizzard to the anus. It is expanded into lateral pouches or diverticula in each somite.

(g) The **chloragogue** cells: the yellowish-brown pigmented layer covering the outside of the stomach-intestine.

Make a diagrammatic sketch (× 4) of the parts included under C. II. 1 to 5 (g).

(h) Open the crop, the gizzard, and the stomach-intestine by a lateral cut.

Examine the character of the wall of each, and note in the stomach-intestine the thick longitudinal ridge, the **typhlosole**, which projects into the intestinal cavity from its dorsal side.

Make a diagram of the stomach-intestine as it would appear in transverse section.

6. The reproductive organs.

[*To see the reproductive organs, cut with fine scissors through the alimentary canal close behind the gizzard, and holding the cut end of the gizzard with fine forceps, raise it carefully and cut through the attachment of the dissepiments (disturbing them as little as possible); cut the alimentary canal across behind the 6th somite and remove the entire œsophagus.*]

In taking out the œsophagus look in the 11th and 12th somites for three pairs of calciferous glands: small pouches attached to the sides of the œsophagus.

Observe:

(a) The **seminal vesicles**: three pairs of relatively large, opaque white bodies lying at the sides of the œsophagus. In the mature worm the posterior pair grow together, forming a **posterior median vesicle** lying ventral to the alimentary canal in the 11th somite; and the two anterior pairs in a similar way form one **anterior median vesicle** in the 10th somite. Cut off the tip of one of the vesicles, mount in water on a slide, and examine with the high power. Look for different stages in the development of the spermatozoa. (Cf. "General Biology," Sedgwick and Wilson, second edition, p. 77.)

(b) The **funnels** ("ciliated rosettes") of the vasa deferentia (seen in the mature worm by carefully opening the median seminal vesicles): two pairs of rosette-like bodies with opaque white borders in the 10th and 11th somites.

(c) The **vasa deferentia**: four fine ducts running posteriorly from the four funnels. The two of the same side unite in the 12th somite to form a single duct on each side, which runs back and opens to the outside in the 15th somite.

(d) The **testes**: four small, opaque white bodies with finger-like lobes, attached one on each side of the median line to the posterior side of the dissepiments 9–10 and 10–11. In the mature specimen they are enclosed by the median seminal vesicles.

(e) The **seminal receptacles**: two pairs of white, rounded sacs in the 9th and 10th somites. Their short ducts open to the outside in the constrictions between the 9th and 10th and the 10th and 11th somites, respectively.

(f) The **ovaries**: two pear-shaped bodies attached to the anterior wall of the 13th somite, one on each side of the median line.

(g) The **oviducts**: two short, funnel-shaped tubes passing through the dissepiment between the 13th and 14th somites, one on each side of the median line. They appear often as thickenings of the dissepiment.

Make an enlarged diagram of the reproductive organs.

7. The nervous system.

Central nervous system. Observe:

(a) On the dorsal side of the pharynx in the 2d or 3d somite a pair of pyriform white bodies united in the median line : the **cerebral ganglia**.

(b) From each lobe a nerve-cord, the **circumœsophageal commissure**, passes down at the side of the pharynx, in the constriction between the buccal sac and the pharynx, to the **subœsophageal ganglion (first ventral ganglion)** on the lower side.

(c) From the subœsophageal ganglion a long double **ventral nerve-cord** passes posteriorly in the mid ventral line.

Peripheral Nervous System.

(d) Two large nerves which run anteriorly from the cerebral ganglia.

(e) A nerve from each half of the circumœsophageal commissure near the union of the two halves on the ventral side.

(f) Typically three pairs of nerves in each somite: two pairs arising from the ganglion and one pair (septal nerves) from the cord immediately behind the dissepiment.

Make a diagram (× 6) of the nervous system. (At this point the buccal sac, C. II. 5. (a), may be looked for.)

8. **The cuticle.**

With forceps strip off under water the cuticle from the lateral and ventral aspects. Mount it in water on a slide, and examine first with a low and then with a high power.

With the low power observe:
  (a) The tubular processes of the cuticle torn from the setæ-sacs.

With the high power observe:
  (b) The thin transparent membrane not composed of cells. It is traversed by delicate lines and perforated by minute pores.

9. Before leaving the anatomy of the earthworm see a demonstration of a nephridium from a recently chloroformed worm.

10. *Make a drawing (× 5) of an egg-capsule.*

D. MICROSCOPIC STRUCTURE OR HISTOLOGY.

I. Prepared transverse section.

For detailed directions see "General Biology," Sedgwick and Wilson, second edition, 1895, pages 91 to 95 inclusive.

Make a diagram showing in outline the different regions of the section.

Sketch accurately, much enlarged, a portion of the section a few cells in width.

Make an accurate sketch of the cross-section of the ventral nerve-cord.

II. Prepared longitudinal sections to show the relations of cerebral ganglia, alimentary canal, circumœsophageal commissure, and ventral nerve-cord.
  1. Section through the median region (cerebral ganglia, alimentary canal, ventral nerve-cord).
  2. Section at one side of the middle region (circumœsophageal commissure).

Diagrams in outline!

AMŒBA.

Mount some of the water or sediment containing amœbæ under a large cover-glass. Search for them with the low power, and when found study with the high power. Observe:

I. General Characters.
   1. Form: in the motile state characterized by irregularity and continual change; rounded processes (**pseudopodia**) formed by the protrusion and the retraction of the protoplasm.
II. Structure.
   1. An outer transparent layer: the **ectoplasm**.
   2. An inner more granular layer: the **entoplasm**.
   3. A spherical or disk-shaped body in the entoplasm: the **nucleus**.
   4. A spherical space filled with fluid, disappearing and reappearing at intervals; the **contractile vacuole**, often situated in the hinder part of the body. If the pulsations can be readily observed, describe them.
   5. **Water-vacuoles** (not always present): in the entoplasm.
   6. Food masses in **food-vacuoles** in the entoplasm.

Sketch an amœba ($\times$ 8 or $\times$ 10).

III. Movements. *Observe:*
1. The mode of formation of a pseudopodium.
2. The process of locomotion.
3. If possible, the taking in of food and the passing out of waste matter.

Make a series of outline sketches at regular intervals to show the changes of form.

4. The effect of higher temperatures on the movements of an amœba. (*Heat gradually on a warm stage.* Record the result.*)

IV. Reproduction.

If an amœba be found dividing by fission, observe and sketch it.

V. Chemical and Mechanical Tests.
1. Treat with magenta or iodine. Record the result.
2. Crush a stained specimen. Describe the result.

---

* A simple warm stage may be made of sheet copper the width of a glass slide and several times its length, with a hole pierced an inch and a half from one end to correspond to the aperture of the diaphragm. To use the stage, place it on the stand of the microscope and fasten the slide down upon it with the microscope-clips. To heat the copper, place a lighted alcohol-lamp under the free end. A simple method of determining when the temperature of the slide has reached approximately 38 degrees is to place upon it a small fragment of paraffine that has been mixed with sufficient benzole to reduce its melting-point to the desired degree.

WHITE BLOOD-CORPUSCLES.

A. *Mount under a cover-glass supported by a piece of hair or a bristle a drop of blood obtained by pricking the finger near the root of the nail with a fine needle. Surround the margin of the cover-glass with vaseline. Make the preparation as quickly as possible in order that it may not dry.*

Disregard the colored corpuscles and look for the less frequent white ones. Observe:

I. General Characters.
  1. Form: irregularly rounded; changes occur as in amœba, but more slowly.
II. Structure.
  1. Granular protoplasm.
  2. Nucleus (rarely visible in the fresh state).
  Sketch.
III. Movements.
  Place the preparation on a warm stage and heat gradually to 38 *degrees C. Record the result.*
B. *Mount in the same way corpuscles of the cœlomic fluid of the earthworm. This may be obtained by exposing the worm for a moment to the vapor of chloroform, when the fluid will exude through the dorsal pores; touch a cover-glass to the fluid and instantly mount.*

Note the shape, and changes of form that may occur.
Sketch.

HÆMATOCOCCUS.

A. RESTING STAGE.*
   Mount on a slide some sediment containing hæmatococcus that has been soaked for a few hours in water. Cover and examine with a low power.
   Find:
I. External Characters.
   1. Red or green spherical cells.
   With a high power observe:
   2. Size. (*Measure.*)
   3. Form.
   4. Color: red, green, or in some cells both red and green.
II. Structure.
   1. Colorless cell-wall.
   2. Cell-contents: more or less granular; consisting of protoplasm, a centrally situated **nucleus** and superficial **chromatophores** in which the coloring matter is deposited. (The outlines of the chromatophores are difficult to distinguish in the living state. The nucleus is more easily seen in the

---

* Hæmatococcus can sometimes be found in old marble urns. When once obtained the same supply will last for a number of years. It should be wet each year with spring-water, and is believed to flourish better if a little powdered marble is added.

smaller, green form, Pleurococcus. See below.)
Sketch ($\times$ 6).
3. Look for cells whose contents are undergoing **endogenous division**. *Sketch.*
4. Treat with iodine. What structures show more clearly?
5. Crush the cells by pressure on the cover-glass. Note the unstained cell-walls, and the cell-contents stained brownish yellow with iodine (showing the presence of protoplasm).

B. Motile Stage: **Zoöspores.**
(These develop from forms in the resting stage that have been in water for a number of hours.)
Study with a high power. Observe:

I. Larger motile forms: **macrozoöspores.**
  1. Form: pear-shaped.
  2. Color.
  3. Structure.
    (*a*) Thin colorless cell-wall.
    (*b*) Central **protoplasmic mass** separated from cell-wall by **clear space.** Through the clear space delicate **protoplasmic threads** extend from the central mass to the inner surface of the cell-wall (probably to a thin layer of protoplasm lining the cell-wall).

    From the **colorless apex** of the protoplasm two **processes** pass to the pointed

end of the cell-wall. They are continued beyond it as rapidly vibrating flagella.
(c) **Nucleus** (not visible in living specimen).
(d) **Chromatophores.**
4. Movements. *Note:*
(a) Active locomotion.
(b) The rotation of an individual on its own axis.
(c) The movements of the flagella. (Try to find a specimen whose movements have become so slow that the flagella may be seen.)
5. Stain with iodine. Look for the flagella. *Sketch* (× 6).
II. Smaller motile forms: **microzoöspores.** Like the macrozoöspores except in regard to size and cell-wall, etc.
Sketch (× 6).

PLEUROCOCCUS.

Soak in water for a few hours pieces of bark, etc., that have a green powdery coating or discoloration. Remove some of the green layer with the point of a scalpel, mount in water, and study with a high power. Look for:
1. **Nucleus.**
2. Individuals multiplying by **cell-division.**
Sketch.

PARAMŒCIUM.

Mount a small drop of water containing Paramœcia under a supported cover-glass. Examine first with AA 2, and observe where they are most quiet; then study with D 2.

I. General Characters.
  1. Form: elongated, flattened, slipper-shaped; the anterior end rounded, the posterior pointed.
  2. Movements: by means of cilia over the entire surface.
  3. On the ventral or oral surface an oblique funnel-shaped depression (**vestibule**) which leads to the **mouth**. Note the direction of the cilia in the vestibule.
  4. A blind sac opening into the vestibule by the mouth; the **œsophagus**.
  5. At a definite point posterior to the blind end of the œsophagus, the **anal spot**, where waste matters are at times passed out from the body.

II. Structure.
  1. **Ectoplasm**: the sharply defined outer layer, with a delicate surrounding membrane, the **cuticle**.

2. **Entoplasm**: the more granular, inner portion.
3. At the base of the vestibule the mouth-opening leads into a blind pouch, the **œsophagus**, through whose posterior wall the food passes into the entoplasm.
4. **Food-vacuoles**: spherical spaces in the entoplasm containing water and food-particles.
5. **Water-vacuoles.**
6. **Contractile vacuoles**: one near the anterior and the other near the posterior end. (*Study carefully, and sketch in two or three stages.*)
7. **Nucleus**: an oval body near the centre of the cell (not always visible in the living state), the **macronucleus**; close to this the **micronucleus.**

Sketch a Paramœcium, much enlarged.

III. Movements.
1. **Currents in the entoplasm**: made evident by changes in the position of the food-vacuoles and the water-vacuoles.
2. **Action of the cilia.** (*To see the currents produced by the cilia in the surrounding water introduce under the cover-glass a small quantity of powdered carmine.*)

Make a diagram of the animal, and indicate by means of arrows the course of the currents.

IV. Reproduction.
Sketch any specimens you may find dividing by fission or conjugating.

V. **Chemical Experiments.**
Treat with 2 per cent. acetic acid; observe: macronucleus, micronucleus, and **trichocysts.**
To see the cilia, treat with dilute iodine.

VORTICELLA.

Mount in water a fragment of a submerged leaf or other substance to which vorticellæ are attached. Examine first with AA 2. Observe form and movements. Look for a quiet specimen and study with D 2. When it is extended, observe:

I. General Characters.
  1. Form : bell-shaped. The bell attached by its smaller end to a stalk.
  2. Parts :
    (a) The conspicuous rim, the **peristome**, surrounding :
    (b) The **disk** : which closes the mouth of the bell.
    (c) Part of the disk raised above the peristome at one side, the **epistome**.
    (d) Cilia : bordering the peristome and the epistome.
    (e) A space between the peristome and the epistome, the **vestibule**.
    (f) The **mouth** : opening from the vestibule into :
    (g) The œsophagus.
    (h) A definite region of the vestibule, the **anal spot**, through which waste matter is passed out from the body (visible only at the time of passing out of waste matter).

II. Structure.
1. The thin transparent external layer, the **cuticle.**
2. The finely granular layer next to the cuticle, the **ectoplasm.**
3. The central portion, the **entoplasm.**
4. **Food-vacuoles.**
5. **Water-vacuoles.**
6. The **contractile vacuole**: generally situated near the disk.
7. The **nucleus**: a long slender curved body (not always visible in the living specimen).
8. The **stalk**: by which the bell is attached, consisting of:
    (a) An outer transparent **sheath** continuous with the cuticle of the bell.
    (b) The contractile **axial filament**, the prolongation of the ectoplasm of the bell.

Make a diagrammatic sketch much enlarged.

III. Movements.
1. Of the cilia.
    Note the way in which the cilia move, the currents produced in the water (*to see this draw a few particles of finely powdered carmine under the cover-glass*), the whirling of the carmine particles down the œsophagus.
2. Of the entoplasm: made evident by the carrying around of food-vacuoles, etc.

3. Of the animal as a whole.
    (a) Method of contraction of stalk and of disk.
    (b) Mode of extension of contracted animal.

Make a diagram of an expanded and of a contracted animal and label corresponding parts.

Make diagrams of stalk (a) in extended state; (b) when contracted.

IV. Reproduction.

Search for :
  1. Specimens dividing by **fission**.
  2. **Motile forms.** When found study with D 2. Observe the secondary band of cilia. Look for the contractile vacuole.
  3. **Conjugation**: the union of a small motile individual with a larger sedentary form.

V. Chemical Experiment.
  1. Stain with iodine ; record the result.

YEAST.

SACCHAROMYCES CEREVISIÆ.

Put upon a slide a drop of liquid that contains actively growing yeast (e.g. brewer's yeast or compressed yeast in Pasteur's fluid), cover and examine with a low power. Note:

I. General Characters.
   1. Aggregations of cells into groups.
   2. In the larger individuals protuberances partially constricted off, **buds**.

Study with a high power, and observe:
   3. Form.
   4. Mode of union.
   5. Size. (*Measure.*)

II. Structure.
   1. Cell-wall.
   2. Cell-contents.
      (*a*) Protoplasm (**cytoplasm**): granular, with shining spherules.
      (*b*) **Nucleus** (not visible in the living state. See demonstration of prepared specimen*).
      (*c*) Vacuoles: note the number and size.

Sketch ($\times$ 10).

---

* For methods of staining, etc., see "General Biology," by Sedgwick and Wilson, second edition, 1895, p. 219.

III. Reproduction.
1. *Sow a few cells of freshly moistened compressed yeast in a few drops of Pasteur's fluid on a slide. Cover, and examine with a high power.* Compare the structure of the cells with that of growing yeast.
Sketch.
Put the slide on a zinc rack, stand the rack in a dish containing half an inch of water, and cover with a bell-jar (moist chamber).
Examine from day to day to see multiplication by budding. (**Gemmation.**)
2. Examine with a high power yeast-cells in which ascospores have been formed. (**Endogenous Division.**)
Sketch.

IV. Chemical Experiments.
1. Mount in water on a slide a very small amount of powdered starch. Look at it with a low power. Draw iodine solution under the cover-glass. Look again; note the effect of iodine solution upon starch.
2. In the same way treat yeast upon the slide with iodine. What inference is to be drawn as to the presence of starch *within* the yeast-cell?
3. Treat another specimen of yeast with magenta solution. What parts of the yeast plant are stained?

4. Crush the cells stained with magenta by pressing forcibly on the cover-glass with the handle of scalpel or needle-holder. What details of structure (see above, II. 1 and 2) are visible after crushing? Does the magenta affect all parts of the cell equally?

V. Physiology.
1. Put a large drop of yeast into each of three test-tubes (previously labelled) containing
    (a) Distilled water;
    (b) Pasteur's fluid without sugar;
    (c) Pasteur's fluid with sugar.
   Stop the tubes with cotton and keep side by side. Examine the tubes from day to day with the naked eye. Record the result and explain.
2. Put a drop of yeast into each of two test-tubes of the same size (previously labelled) containing equal amounts of Pasteur's fluid with sugar. Stop both well with cotton. Boil one for five minutes. Keep them side by side. Examine the tubes with the naked eye from day to day. Record the result and the explanation.

PENICILLIUM.

Observe Penicillium growing upon bread in a moist atmosphere.* *Note:*

I. General Characters.

Velvety-looking surface, — white in young specimens, — dull bluish green in older stages.

Touch a mass of Penicillium spores with a needle-point, and dip the needle-point into a drop of water upon a slide. Cover, and examine with a low and then with a high power. Note:
Size. (*Measure.*)
Form.
In the same way sow some spores in Pasteur's fluid in a watch-glass. Put the watch-glass into a moist chamber. With a hand-lens watch the development of the **mycelium** *from day to day.*

II. Structure.

Examine germinating spores (sown twenty-four hours previously in a vessel of Pasteur's fluid).
Observe and sketch a series of stages in the germination of the spores, noting:

---

* To obtain Penicillium, keep a cut lemon for a few days in a moist atmosphere.

1. Formation of **hyphæ**.
    (*a*) Cell-wall.
    (*b*) Protoplasm.
    (*c*) Vacuoles.
    (*d*) Fat-drops.
Mount in water a portion of mycelium from bread. Observe:
2. Hyphæ.
    (*a*) Division into cells.
    (*b*) Mode of branching. From what part of a cell does a branch arise?
    (*c*) Vacuoles.
Sketch.

III. Reproduction.
   Aerial hyphæ and conidiophores.
Tease out in water some of the mycelium from a specimen that has become only slightly green. Observe:
1. **Erect hyphæ**, each consisting of:
    (*a*) Primary erect hypha.
    (*b*) Branches. (Terminal branches: **basidia** or **sterigmata**.)
    (*c*) Constrictions of the terminal branches: spores (**conidia**). At which end of the chain are the conidia larger? At which end are they first formed?
Sketch.

LICHEN.

I. External Characters.

Examine with the naked eye the thallus of a green foliaceous lichen—e.g., Parmelia—that has been moistened with water for several hours. Observe:

1. Form.
2. Color.

Sketch a few lobes.

With a scalpel carefully separate part of a thallus from the substratum. Observe:

3. The difference in color between the two sides.

Place the specimen uncovered on a slide, the lower surface uppermost, and examine with AA 2 by reflected light. Observe:

4. Rhizoids. (Many will have been broken off and only their bases will show.)

II. Structure.

1. Sections.

With a razor make sections through the thick part of a thallus; or, the specimen may be rolled up closely and cross-sections of the roll may be made.

Mount (in water) so as to show the cut surface. Examine with AA 2 in strong reflected light Observe:

(a) A thin white superficial layer: the **upper cortical** layer.

(b) A green layer: **gonidial** layer.

(c) A thick, pure white layer: **medullary** layer.

(d) A dark-colored, external layer: **lower cortical** layer.

Sketch.

2. *Place a small piece of the thallus in water on a slide and tear it apart with needles as completely as possible. Cover and examine with D 2. Observe:*

(a) The hyphæ. Disregard the broken fragments and search for specimens in their natural condition.

(b) The **gonidia**. Observe: .

Cell wall.

Chlorophyll.

Nucleus (not always visible).

Sketch a few hyphæ and gonidia to show their form, their relative size, and the manner in which they are connected.

MUSHROOM.

I. External Characters.
   Note:
   1. The vertical stalk, or **stipe**.
   2. The umbrella-like cap, the **pileus**.
   3. The ring, or **annulus**, around the **stalk**, the remains of the membrane, the **velum**, connecting the pileus with the stipe and torn by the extension of the pileus.
   4. The underground mycelium from which the stipe arises.
   5. On the lower surface of the pileus radiating vertical plates or **lamellæ**.
   Sketch.
II. Structure.
   Tease out in water a piece of the stalk.
   1. Note the hyphæ of which it is composed.
   Sketch.
   Examine under the low power the surface of a lamella.
   2. Note the spores; sometimes in groups of four.
   Cut in pith cross-sections of a lamella.
   Note:
   3. The centre of the lamella, made up of hyphæ running parallel to the surface.

4. The surface of the lamella consisting of hyphæ at right angles to the surface.
5. The swollen ends of many of the hyphæ bear four small awl-like points, the **sterigmata**, each with a single spore.

Sketch.

Tease out in water some of the underground mycelium.

6. Note the hyphæ.

Sketch.

BACTERIA.

PRELIMINARY.

Mix some India ink or finely powdered gamboge with water and examine a drop with a high power.

Note that when the currents in the water have ceased the particles of gamboge do not move from place to place.

Now observe the vibration or oscillation of the lifeless particles (**Brownian movement**).

BACTERIA.

I. Active Bacteria.

Examine with the highest power a drop of hay infusion or of other liquid containing active bacteria. Observe:

1. Form.
2. Size. (*Measure*).
3. Structure. (*What points of structure are visible?*)
4. Movements.
    (*a*) Brownian movement.
    (*b*) Active movement from place to place.

Treat with iodine.

Does the iodine bring out any details of structure. What is its effect upon the movements of the bacteria?

46 GENERAL BIOLOGY.

II. Resting Bacteria—Zoöglæa Stage.
Examine the scum (zoöglœa) from the surface of hay or of other infusions, or of aquaria, etc. Observe:
The multitudes of motionless bacteria imbedded in a gelatinous substance.

III. Cover-glass Preparation.
With the forceps hold a perfectly clean cover-glass nearly horizontal and bring its lower side just into contact with the surface of the liquid containing bacteria (the best results are from a scum that is only just visible). Put the cover-glass, wet side uppermost, on a piece of filter-paper slightly inclined in order that the water may drain off. When the cover-glass is dry, pass it quickly three times through a flame; put a drop of methyl violet upon the cover-glass; after three or four minutes rinse off with distilled water; invert the cover-glass upon a drop of water on a slide and examine with a high power. Observe:
The deeply stained bacteria.
The differences in form and size.
Sketch a few of each form.

IV. Experiments.
1. *Clean a potato thoroughly, sterilize it in steam for an hour, and cut it lengthwise into halves with a knife sterilized in the flame of a Bunsen burner. Put the halves on a*

BACTERIA.

sterilized glass plate and expose the cut surface to the air for an hour. At the end of this time cover with a bell-jar that has been sterilized, and put under this some (sterilized) liquid to make a moist atmosphere.

Examine from day to day for ten days or two weeks. Record results and explanation.

2. *Put some freshly-made hay infusion into three (previously labelled) test-tubes (a), (b), (c). Stop them securely with cotton.*

   (a) *To be set aside.*
   (b) *To be boiled once for 3–5 minutes.*
   (c) *To be boiled for 3–5 minutes three or four times at intervals of nine or ten hours.*

Keep all three side by side. Watch carefully for any changes that may appear. Describe briefly, with explanation.

SPIROGYRA.*

I. General Characters.
Observe with the naked eye. Note:
1. Floating masses consisting of long, fine, green threads.

Take between the fingers and note the smooth, slippery feeling.

Mount a small quantity, in water on a slide, cover, and examine with a low power. Note:
2. Unbranched filaments.
   (a) The shape of the cells.
   (b) The way in which they are joined.
   (c) The difference between the terminal cell and the other cells of the filament.

With a high power observe:
3. The form of the cells composing the filament. Focus carefully and determine what geometrical form the cell most resembles.

II. Structure.
With D 2 examine:
1. The cell-wall. (Which part is common to two cells?)
2. The protoplasm.

---

*Spirogyra can often be found in slowly-running streams or floating in the water of quiet pools.

(a) Thin colorless layer lining the cell-wall, **primordial utricle**.

To see the primordial utricle better, treat with 5 per cent. or 10 per cent. salt solution. Watch carefully and observe:

The contraction of the primordial utricle away from the cell-wall (**Plasmolysis**). *Sketch.*

With a fresh specimen observe:

(b) One or more green bands, generally with jagged edges, imbedded in the primordial utricle: the **chromatophore**.

Changing the focus slowly, follow the band from one end of the cell to the other.

What is its length compared with that of the cell? How is it arranged?

(c) Biconvex disk-shaped bodies in the chromatophore: **pyrenoids**.

(d) Near the centre of the cell a small mass of protoplasm suspended by fine threads that connect it with the protoplasm lining the cell-wall; in its centre is the **nucleus**, with a large **nucleolus** (not always visible in the unstained specimen).

3. Enclosed by the primordial utricle and occupying most of the cell-cavity: a large **vacuole**.

50   GENERAL BIOLOGY.

Sketch a complete cell with half of each adjacent cell (× 5).

Stain slightly with iodine and carefully re-examine the cell-structures.

*Look again at the different structures of the cell in preserved specimens stained with borax carmine.**

I. Reproduction.

Mount in water a group of filaments of conjugating spirogyra. Examine with D 2.

[There may be present several species of Spirogyra—differing in the number of the chromatophores. In addition to Spirogyra other closely related forms may also be present,—as Mesocarpus and Zygnema,—differing chiefly in the size of the filament and in the shape of the chromatophore.]

Search for a series of forms to show the different stages in the process of conjugation.

Observe the **zygospores**: large and oval. [Those of Zygnema smaller and more rounded.]

In perfect specimens each filament containing zygospores is connected with a filament with empty compartments. Find cells whose walls have just begun to send out a projection from

---

* To preserve Spirogyra, put it into a saturated aqueous solution of picric acid for 12 hours. Wash for an equal length of time in water. Stain for some hours in borax carmine.

the side. What is the relation between the condition of the cell-contents and the presence or absence of connecting tubules between two filaments?

Sketch a series of stages.

HYDRA.

I. General Characters.
   Look at hydras in an aquarium. Observe that they are found in greatest number on the illuminated side. Notice:
   1. Form.
   2. Color.
   3. Movements.
   With a glass tube transfer one or more specimens to a watch-glass; place this on white paper and examine the hydra with a hand-lens. Observe:
   4. The **body**: cylindrical and varying in length and thickness with the degree of extension of the hydra.
      The parts of the body:
      (a) The basal part, the **foot**, by which the animal adheres to foreign surfaces.
      (b) The conical free end, the **hypostome**.
      (c) The opening at the summit of the hypostome, the **mouth**.
   5. Highly contractile processes of the body-wall arranged around the mouth, the **tentacles**. Count them.
      Note the symmetrical arrangement of the tentacles around the mouth—an instance of

radial symmetry (the arrangement of "like parts about an axis from which they radiate").

6. Buds. (Often present.)

Make simple outline sketches of the hydra when extended and when contracted.

II. Structure.

Transfer by means of a pipette to a slide and mount in water, supporting the cover-glass at each side by another cover-glass. Study with AA 2.

Review the parts examined with a hand-lens. Observe also:

1. The transparent colorless outer layer, the **ectoderm**.
2. The inner thicker layer, the **entoderm*** (green in H. viridis).

With D 2 examine the character of the ectoderm (best seen in extended tentacles). Find:

3. The knob-like prominences, the **nettle-batteries**,† consisting of numerous **cnidoblasts** (parent-cells of **nematocysts**), each with a stiff projecting process, **cnidocil**.

Each cnidoblast contains a highly refractive body enclosing a thread (nematocyst).

Kinds of nematocysts:

---

* The green hydra shows best the distinction between ectodermal and entodermal cells, especially when macerated.

† Best seen in the brown hydra.

(a) Large, often occupying the centre of the nettle-battery.

(b) Smaller and more numerous, surrounding (a).

When a tentacle is in focus under AA 2, draw a few drops of magenta under the cover-glass. Observe:

The discharge of the threads.

With D 2 find the threads, and also note in the surrounding water discharged nematocysts.

(a) The larger rounded cysts do not become colored.

Note:

The **barbs.**

The long finely tapering thread.

(b) The smaller oval nematocysts stain deeply. Note:

The shorter, thicker thread.

Sketch both kinds of nematocysts.

III. Histology.

A. Isolated Cells.

Tease out in dilute glycerine a hydra that has been macerated and slightly stained.* Examine the fragments with the highest power in a strong

---

* To macerate, put for a few minutes into a mixture of glacial acetic acid 1 part, glycerine 1 part, water 2–4 parts (the method of Bela Haller), and stain in an aqueous solution of methyl-green.

light. Search for the following kinds of cells, and sketch those clearly observed:

1. Entoderm-cells, of very transparent protoplasm, containing in H. viridis numbers of **Zoöchlorella** cells.
2. Interstitial cells: small angular granular cells with distinct nuclei:
3. Cnidoblasts containing nematocysts; look for the cnidocil and the nucleus.
4. Large ectoderm-cells: pale cells with large nuclei; the bases drawn out into muscular processes.
5. Nerve-cells: small granular cells with long branches and distinct nuclei.
6. Gland-cells (entodermal): columnar cells with distinct nuclei and foam-like, granular protoplasm.

B. Prepared Sections.*

With AA 2 observe in transverse section:
1. General form.
2. Ectoderm.
3. Entoderm.
4. **Supporting lamella.**

---

* For sections the hydras may be immobilized in a .25 per cent. aqueous solution of hydroxylamin in from fifteen to sixty minutes, and then preserved in picro-acetic acid; or they may be killed instantly in the extended state in a few drops of water in a watch-glass by pouring upon them hot, saturated aqueous solution of corrosive sublimate. Fine preparations may also be made from hydras preserved in osmic acid.

Make an outline sketch of the section, omitting cell-details.

With D 2 study the cell-structures as far as they can be made out.

In the ectoderm:

(a) Large vacuolated ectoderm-cells; the nuclei can be seen in some sections.

(b) Between the large ectoderm-cells rows of smaller deeply stained cells, the **interstitial cells.** Many of these are developing into:

(c) Cnidoblasts.

(d) A row of minute bodies like dots lying along the outer side of the supporting lamella: sections of the **contractile processes** of the large ectoderm-cells.

Between ectoderm and entoderm the thin, structureless, supporting lamella.

In the entoderm:

(e) A single layer of large, much vacuolated cells. (What is the explanation of the appearance in the section of several layers of cells?)

(f) Small gland-cells among the ordinary entodermal cells.

Sketch accurately part of the section a few cells in width.

Longitudinal Section.

Note the central cavity enclosed by a wall consisting of two layers separated by a lamella.

Make a diagrammatic sketch.

Find again the cells seen in transverse section.

FRESH-WATER MUSSEL.*
ANODON.

A. General Characters.
I. Bivalve shell enclosing the animal, the valves fastened together by a hinge along one—the dorsal—side.
  1. Straight dorsal side.
  2. Curved ventral side.
  3. Anterior end wide and rounded.
  4. Posterior end narrower and more pointed.
  5. The bivalve shell bilaterally symmetrical.
  6. Concentric lines, **lines of growth** of the shell.

Diagram!
Force apart the valves of the shell about half an inch and insert a wedge. Find:
  7. The **mantle,** lining each valve of the shell.

With the handle of a scalpel separate the mantle from the right valve. Find:
  8. The **adductor** muscles.

With a strong scalpel cut through the attachment of the muscles on the right side close to the

---

* Living mussels are advertised by some of the dealers in laboratory supplies. For preservation they should be put for twenty-four hours into 70 per cent. alcohol (which is reduced to a much lower per cent. by the water inside the shell), then changed to 70 per cent. alcohol for twenty-four hours, and kept permanently in 80 per cent. alcohol.

shell. Remove the right valve. On its inner surface find:
9. The line of attachment of mantle to shell.
10. Impressions of muscle-attachments.
    Near the anterior end:
    (a) A large impression made by the **anterior adductor** and the **anterior retractor** muscles.
    (b) A small impression near (a), that of the **protractor** muscle.
    Near the posterior end:
    (c) A large impression, of the **posterior adductor** and the **posterior retractor** muscles.
As the mussel lies on the left valve, find:
11. The muscles whose impressions have been seen in 10. *Observe:*
12. The mantle enclosing the other parts of the animal.
13. The edges of the mantle closely applied except at the posterior extremity, where they separate to leave two orifices:
    (a) A smaller dorsal opening (through which water passes out), the **exhalent aperture**.
    (b) A larger opening more ventral (through which water enters), the **inhalent aperture**.
Diagram!
Insert a bristle into each opening, and then raising the flap of the mantle see:
14. Inserted into the more ventral opening it

passes into the space between the flaps of the mantle, the **pallial** or **mantle-cavity**.

15. Inserted into the dorsal opening it enters the **cloacal chamber** and passing forward enters the mantle-cavity.

In the mantle-cavity find the following structures:

16. Inside of the mantle on each side two gills or **branchiæ**.
17. Projecting ventrally near the anterior end, the thick **foot**, continued dorsally as the **visceral mass**.
18. Attached to the mantle anterior to the gills two triangular folds on each side, the **labial palps**.

The gills, the foot, and the labial palps all hang down in the mantle-cavity. By the union of the inner gills in the posterior median line a partition is formed that divides the posterior part of the mantle-cavity into two:

The large ventral **branchial chamber**.

The smaller dorsal **suprabranchial chamber**.

The posterior end of this forms the cloacal chamber.

Pass a bristle again into the cloacal chamber, through the suprabranchial chamber, and into the branchial chamber.

II. External Openings.
1. The labial palps are united at their bases to form ridge-like anterior and posterior lips. Between these is the mouth.
2. On either side of the visceral mass near the attachment of the gills is a small genital opening.
3. Close to 2, slightly more dorsal, is the excretory opening.
4. On the dorsal side of the cloacal chamber close to its external opening is the anus.

Make a diagram of the animal as it would be seen if the mantle-flap of one side were absent.

B. Anatomy.
I. Circulatory System.

On the dorsal side of the body immediately in front of the posterior adductor muscle is the pericardial cavity with thin semi-transparent walls.

Cut with fine scissors through the thin wall a little to the right of the median dorsal line, taking care not to cut deeper, thus exposing :

1. The pericardial cavity. It contains :
2. The heart.
   (a) In the median line the opaque, muscular ventricle.
   (b) Into the ventricle opens on each side a thin-walled auricle. The auricles are funnel-shaped with the narrow end at-

tached to the ventricle and the broader end to the dorsal border of the gills.
3. The vena cava running parallel to the heart under the floor of the pericardial cavity; it will be seen later (C. II.) in section.

Note the rectum passing posteriorly from the posterior end of the heart.

Make a diagram of the heart.

II. Excretory System.

In the anterior angle of the pericardial cavity on each side is a minute pore, the internal opening of the excretory organ, the **organ of Bojanus**, the dark mass on each side beneath the floor of the pericardial cavity, (For the external opening of the excretory organ see A. II. 3.)

III. Nervous System.

Remove the mussel from the shell, place it dorsal side down, and cut through the partition between the branchial and the cloacal chambers. See:

1. Close to the ventral side of the posterior adductor, the **visceral or parieto-splanchnic** ganglion.

Gently strip off the thin skin covering the ganglion, and find the chief nerves given off from it:

(*a*) The **posterior pallial** nerve.
(*b*) The **lateral pallial** nerve.
(*c*) The **branchial** nerve.

(d) A nerve passing forward on each side of the middle line, the **cerebro-splanchnic commissure.** Trace one forward through the body to the :
2. **Cerebral** ganglion of that side, lying close beneath the skin at the base of the labial palps immediately in front of the protractor muscle.

Find the cerebral ganglion of the other side, and the commissure connecting the two across the mid-line around the dorsal side of the mouth.

3. Deeply-buried in the foot along the plane of its junction with the visceral mass, the **pedal** ganglia, closely connected by a transverse commissure.

They may be found by making with a razor a median vertical section of the foot, or by following the commissure from the cerebral ganglion of one side.

Find the commissures connecting pedal and cerebral ganglia.

Make a diagram of the nervous system.

IV. The Digestive System.

Insert a small blunt instrument into the mouth. With this as a guide open the alimentary canal with fine scissors. Observe:

1. The mouth.

2. The œsophagus : a short straight tube extending obliquely upward to :
3. The stomach : a dilatation of the alimentary canal.
(The intestine passes with many turns through the visceral mass, and will be seen later in sections.)
In the same way open the rectum from the anus, and trace its course through the pericardial cavity.
Look for the **typhlosole**.
V. The Gills.
1. Take out one of the gills, cut it transversely, and examine the cut edge with a hand-lens.
2. Examine with the low and then with the high power a small piece of one lamella.
3. In some specimens the outer gills are distended with young mussels (**glochidia**). If these can be obtained, mount and examine with a low and with a high power.
C. Sections.
Remove the shell from a second specimen, place the mussel on its side upon the wax in a dissecting-pan, and with a razor make sections a quarter of an inch thick. Arrange the sections in order, fastening each in place with a pin, and cover them with water.
Study especially the following sections:

I. Through the pedal ganglia.
   Find the stomach, digestive gland, labial palps, foot, and mantle.
II. Through the heart.
   Find the ventricle, the rectum with the typhlosole, auricles, organ of Bojanus on each side with the vena cava between, the visceral mass with the intestine cut through in a number of places, the reproductive organ filling up the greater part of the visceral mass, the foot, the inner and outer gills, the suprabranchial chamber dorsal to the outer gill on each side, the branchial chamber, the mantle.
III. Through the visceral ganglia.
   Find the rectum, the posterior adductor muscle, the visceral ganglia, the divisions of the suprabranchial chamber (how many?), the branchial chamber, the mantle.

Diagram of these three sections!

In the other sections trace the organs from section to section.

LOBSTER.*

A. EXTERNAL CHARACTERS.
I. The hardened cuticle, the **exoskeleton**, covering the animal.
II. The body, consisting of:
  1. An anterior part, the **cephalothorax**, with a firm continuous covering above (**carapace**).
  Parts of carapace:
   (a) Cephalic part, covering the head of the lobster; prolonged anteriorly into the **frontal spine**, on each side of which is a depression containing the stalked eye; limited posteriorly by the **cervical groove** which separates it from:
   (b) Thoracic part, covering the thorax, consisting of:
      (aa) Median dorsal part, separated by a very slight groove on each side, the **branchio-cardiac groove**, from:
      (bb) Two lateral parts, one on each side, the **branchiostegites**, which protect the gills.

---

* If lobsters cannot be obtained in the market, these directions will answer nearly as well for the crayfish.

Raise the branchiostegites and see beneath them the gills.
2. A jointed flexible posterior part, the abdomen.
Count the number of segments. In what plane or planes can the abdomen be moved?
III. Appendages.
Jointed limbs attached to the ventral aspect of the body. (To be studied in detail later (C. II.)).
IV. External Openings.
1. The slit-like auditory openings: on the flat upper surface of the basal segments of the antennules.
2. The excretory openings, perforating a tubercle on the ventral side of the basal segment of each antenna.
3. The mouth, in the median line, seen by pushing apart the appendages on the ventral side of the head.
4. The genital openings:
  (a) In the female: on the basal segment (**coxopodite**) of the second from the last thoracic appendage.
  (b) In the male: on the basal segment of the posterior thoracic appendage.
5. The anus: a slit on the ventral surface of the posterior abdominal segment.

B. ANATOMY.
I. Circulatory System.

Cut with strong scissors through the two branchio-cardiac grooves and transversely through the cervical groove (taking great care not to cut through the skin beneath) and remove the isolated part of the carapace.

Note:

The skin, colored by red pigment.

Cut through the body-wall by a shallow cut in the median line and a transverse cut at each end of the median cut. Turn aside the flaps, thus opening the **pericardial sinus** *and exposing:*

1. The heart: pentagonal and slightly yellowish, lying in the median line.

    Find delicate strands of muscle that pass from the angles of the heart to the wall of the pericardial sinus.

    The heart receives blood from the pericardial sinus through three pairs of valve-like openings, **ostia**:

    (a) A dorsal pair: visible now.

    (b) A lateral pair: better seen at a later stage (B. I. 2. (e)).

    (c) A ventral pair: better seen at a later stage.

2. Arteries: arising from the heart. (These are extremely delicate and must be looked for with great care.)

    From the anterior region of the heart:

(a) A median **ophthalmic** artery.

(b) A pair of **antennary** arteries, close to (a), one on each side.

(c) Posterior and ventral to (b), a pair of **hepatic** arteries.

From the posterior region of the heart:

(d) A large median **superior abdominal** artery. By longitudinal cuts remove a narrow piece of the abdominal exoskeleton and follow the artery through the abdomen.

(e) Passing ventrally from the ventral aspect of the heart a large median **sternal** artery: better seen at a later stage. (See below.)

Diagram of heart and of arteries thus far seen.

Follow out the five arteries from the anterior end of the heart, as far as possible without injury to the other organs.

Cut through the anterior arteries, and turning back the heart find the lateral and ventral ostia (B. I. 1. (b), (c)), and the sternal artery (B. I. 2. (e)).

Cut through the sternal artery close to the heart and through the superior abdominal artery and take out the heart.

Finish diagram of the heart and arteries!

II. Reproductive System.

On each side of the heart and partly ventral to it, find:

In the male:

1. The testes: two long, white, lobed bodies which extend posteriorly into the abdomen and are connected near their anterior ends by a transverse commissure.

    Near the middle of each testis a vas deferens arises and passes ventrally to the genital opening (A. IV. 4. (b).)

By cutting through the muscles, etc., at the side, follow the vas deferens of one side from the testis to the external opening.

In the female:

2. The ovaries: elongated, yellowish bodies continued posteriorly into the abdomen, with the oviduct passing to the genital opening (A. IV. 4. (a)).

Trace the oviduct from the ovary to the external opening.

Diagram of reproductive organs and ducts!

III. Digestive System.

In the median line anterior to the region of the heart:

1. The stomach a dilatation of the alimentary canal.

Insert the handle of a section-lifter through the mouth-opening into the stomach. Turn the stomach to one side to see:

2. The œsophagus: a short vertical tube from mouth to stomach.

LOBSTER.

Diagram of the œsophagus and stomach as they would appear if seen from the side. Observe:

3. Posterior to the stomach on each side a large, soft, brown or greenish-yellow structure, the **digestive gland.**

Gently pushing it with the handle of a scalpel away from the middle line, find:

4. The duct of the digestive gland of that side, a short, wide, thin-walled tube which enters the part of the alimentary canal, the **mesenteron**, immediately posterior to the stomach.

Removing the reproductive organs and the digestive gland, and separating the muscles of the abdomen in the median line, find:

5. The intestine : a straight tube leading from the mesenteron to the anus.

Finish the diagram of the digestive system. Cut with fine scissors through the œsophagus close to the stomach, and through the intestine, and take out the stomach. Open the stomach and examine:

6. The arrangement of the **gastric teeth.**

IV. Excretory System.

In the extreme anterior part of the cephalic cavity on each side.

1. Paired, delicate, greenish structures, the excretory organs (**green glands**).

(For the external openings see A. IV. 2.)

V. Endophragmal System.

Calcified projections from the inner side of the exoskeleton of the ventral surface of the thorax form an internal skeleton—the **endophragmal skeleton**, which arches over a canal —the **sternal canal.**

To see the endophragmal system remove the thoracic viscera and the muscles ventral to them.

VI. Nervous System.

Cut through the endophragmal system at each side of the median line, thus exposing the thoracic part of:

The central nervous system, which runs the whole length of the body, in the median line, consisting of :

1. The supraœsophageal or cerebral ganglia, dorsal to the œsophagus. Each ganglion gives off nerves to the eye, to the antennule, and to the antenna of its own side.
2. The circumœsophageal commissure passing posteriorly and ventrally from the cerebral ganglia around the œsophagus to the first pair of :
3. The thoracic ganglia : six pairs united by longitudinal commissures.

Follow the commissures passing posteriorly from the posterior pair of thoracic ganglia, and find in the abdomen :

4. The abdominal ganglia: six pairs likewise connected by longitudinal commissures.

Diagram of the nervous system !

C. EXOSKELETON AND APPENDAGES IN DETAIL.

I. Typical Abdominal Segment.

Make a diagram of the third abdominal segment as it would appear if seen in anterior or posterior view, to show the form of the segment and its appendages.

II. Appendages.

Take off all the appendages from one side of the body, beginning at the posterior end, being especially careful to remove each appendage entire.

Observe:

The appendages that bear the gills.

The gills that are borne elsewhere.

Compare each different kind of appendage with the second maxilliped, and by reference to the chart * determine the homologous parts in the series.

Find the **scaphognathite.** Of what parts is it made up? In what way can it move?

Of what parts is the **telson** composed?

Diagrams of :

The third abdominal appendage.

---

* Enlarged from Plate VIII, The Appendages of the Crayfish, in "The Atlas of Practical and Experimental Biology," G. B. Howes.

The posterior thoracic appendage.
The fifth thoracic appendage.
The fourth thoracic appendage.
The second maxilliped.
The first maxilliped.
The second maxilla.
The first maxilla.
The mandible.
The antenna.
The antennule.

MOSS.
POLYTRICHUM.

A. SEXUAL GENERATION. (Oöphore.)
Keeping the specimen moist, observe:
I. General Characters.
  1. Stem: erect and unbranched (or branching only from the base).
  2. Leaves: pointed at one end. Attached to the stem without a stalk (i.e., sessile).
  3. Rhizoids: at the base of the stem (often hidden by the sand, etc., clinging to them; to see them distinctly, rinse in water).
  4. Flowering heads.
     (*a*) In male flowers a terminal rosette of stiff green leaves (**perichætium**) surrounding the antheridia.
     (*b*) In female flowers a terminal bud formed by the folding together of terminal leaves that enclose the archegonia.
  Sketch.
II. Histology.
  1. Stem.
  Remove the leaves near the base of the stem; with a razor make a thin transverse section, mount in water, and observe:
     (*a*) The thin epidermis (best seen in young

specimens), and the subepidermis: a conspicuous brown layer.

(b) Next to this a region of nearly colorless cells. Compare these with (a) in regard to cell-contents and thickness of cell-wall.

(c) In the centre a central or axial strand.

Make a diagram showing the outlines of the different regions. Sketch a few cells of each kind to show their characteristics.

Make a longitudinal section, mount in water and find the same regions. Sketch carefully to the same scale a few cells of each kind.

2. Leaf.

Mount in water a young leaf of a more delicate species of moss—as Mnium—and study with D 2 to see:

(a) The structure, a thin lamina with a thicker median region, the midrib.

(b) The margin of the lamina.

Sketch two or three cells at the edge of the leaf to show the character of the margin.

3. Rhizoids.

Mount in water. With D 2 note:

(a) Cell-contents.

(b) Any difference that is observable between the younger and the older rhizoids.

Sketch.

4. Reproduction.

A. Male Plant. *Sketch.*
With a razor make longitudinal sections of the flowering head. Mount in water and observe at the apex of the axis within the perichœtial leaves:
(a) Antheridia : oblong sacs with a wall of a single layer of cells. Within the sacs are mother-cells of antherozoids.
(b) Filaments (**paraphyses**).
  (aa) Hairlike.
  (bb) Spatula-shaped.
Sketch an antheridium from the surface and in optical section.
Sketch each kind of paraphysis.
B. Female Plant. *Sketch.*
With needles separate the leaves from the top of the axis of the female flowering head and search for:
(c) Archegonia : flask-shaped bodies with elongated neck and enlarged ventral part containing the oösphere.
(d) Paraphyses.
Sketch.
B. ASEXUAL GENERATION. (Sporophore.)
Observe:
I. General Characters.
1. The slender stalk, **seta,** which supports :
2. The **capsule,** which contains the spores.
Sketch.
The parts of the capsule :
(a) The thick hairy cap, **calyptra.**

Take off the calyptra and by reflected light note beneath it:
(b) The **operculum.**
Sketch the capsule with the calyptra removed. Take off the operculum and observe:
   (c) Around the rim on which it rested, the **teeth** of the **peristome.**
   (d) Stretched across between the tips of the teeth, the **epiphragma** partially closing the spore-case.
Divide the capsule transversely and make a diagram of the cut surface.
Examine and sketch a spore.

C. PROTONEMA.

Examine with the naked eye some living moss-protonema.*
Mount some of the protonema in water; with needles gently separate the green filaments; cover and examine with D 2. Observe:
1. The green branching filaments made up of cells.

Sketch.

2. Growing out in places from the protonema **buds** developing into moss-plants.

If observed, sketch.

---

*Found growing as a green encrustation on banks of earth without much other vegetation, by roadsides, etc., in partly shaded places, especially those exposed to the north.

SPERMAPHYTES.
STEM.

I. Gymnosperm.
Make with a razor thin sections—transverse, radial, and tangential—of pine wood that has been soaked in water. Mount in dilute glycerine and study with AA 2 and D 2.
Observe in the transverse section :
1. The **annual rings**.
Compare the two edges of a ring.
2. The thick-walled pitted tracheids.
3. The narrow medullary rays.
Make a diagram with AA 2 showing the relation of the autumn to the spring wood, and of the wood to the medullary rays.
With D 2 sketch accurately a small part of the section.
In the tangential section observe :
1. The tracheids : their form, their arrangement, and the pits in their walls.
2. The medullary rays.
Make a diagram showing the relation of tracheids and medullary rays.
Sketch accurately with D 2 a small part of the section.
Observe in the radial section :

1. The tracheids. Why is a larger number of pits visible in this section?
2. The medullary rays.

Make a diagram showing the relation of tracheids and medullary rays.

Sketch a few tracheids much enlarged.

From what has been seen in the three sections describe the shape of a medullary ray.

II. Angiosperm.
A. Dicotyledon.

Ampelopsis. Prepared transverse section of entire, young stem. *Study with AA 2 and D 2.*

AA 2. Observe:
1. Epidermis, one cell-row in thickness.
2. Cork (subepidermis) consisting of several layers of flattened cells.
3. Cortical parenchyma.
4. Fibrovascular bundles.
5. Pith.
6. Medullary rays.

Diagram.

Which tissues have cell-contents?

Cut a section from fresh material and see which tissues contain chlorophyll.

D 2. In the fibrovascular bundle observe the **cambium** layer dividing the bundle into an inner part (**xylem**) and an outer part (**phloëm**).

Xylem, consisting of:
7. Large vessels or tracheæ.
8. Tracheids, smaller in cross-section than the vessels.
9. Wood parenchyma (?). Thin-walled cells with cell-contents.

Phloëm, consisting of:
10. Sieve-tubes, thin-walled and in some places showing the perforated ends of the cells.
11. Bast-cells, thick-walled.

Sketch accurately a few cells of each kind in their relation to one another.

B. Monocotyledon.

Observe with the naked eye the cut end of a stem of Indian corn. Note:
1. The firm outer layer (epidermis and sub-epidermis).
2. The fundamental parenchyma.
3. The isolated fibrovascular bundles.

Make thin cross-sections through the outer part of the stem. Mount in dilute glycerine and study with AA 2 and D 2.

Observe in the bundles:
4. The two large pitted tracheids.
5. The smaller, spiral and annular vessels (one or more) between the tracheids and nearer the centre of the stem. In many instances the cells immediately surrounding these

vessels have disappeared, leaving an irregular space.
6. The thin-walled phloëm on the side opposite the vessels.

The prosenchyma forming the outer part of the bundle does not belong to the bundle itself. It is a bundle-sheath of modified fundamental tissue.

Root.

Lay in water on a slide the root of a young mustard-seedling grown in sawdust. Cover and examine with AA 2.
1. Find the region in which root-hairs are present.

Sketch.

2. Moisten a piece of blue litmus-paper and touch it to the root of a fresh specimen. Record the result.
3. Examine the root-cap of Pontederia or of Lemna.

Sketch under AA 2.

Bud.

Winter buds of elder or of horse-chestnut.
1. *Sketch the bud.*
II. *Take off the bud-scales and the undeveloped leaves in regular order, beginning with the outermost.*
Notice the way in which they overlap.

III. Compare as to texture the outer scales with what is inside. Suggest a reason for the difference. What other means are there for the same purpose? (Make a note of the answers.)

IV. Compare a closed bud with others * in different stages of unfolding. Make a note of what is observed.

LEAF.

I. *Strip off some of the epidermis from a hyacinth-leaf. Mount in water and examine with AA 2 and D 2.*
Observe:
1. The elongated cells of the epidermis without chlorophyll. Compare with the epidermis of the fern.
2. The stomata, each with two guard-cells containing chlorophyll.

Sketch a few cells accurately (D 2).

II. *Cut a transverse section of the leaf of the india-rubber tree* (Ficus elastica). *Mount in water. Draw accurately.*

Beginning with the upper surface find:
1. The upper epidermis, consisting of several

---

* Buds in different stages of opening can be obtained at any time during the winter by keeping cut stems in water for a few weeks. Flowering branches with the flower in different stages of unfolding should be preserved in alcohol from the previous season.

layers of thick-walled cells without chlorophyll. Compare with the epidermis of the fern.
2. Elongated cells closely packed together, with their long diameter at right angles to the surface of the leaf, the **palisade** tissue, containing many chlorophyll-bodies.
3. Other cells containing chlorophyll, somewhat irregular in shape and loosely arranged, so that between them are large open spaces, **intercellular spaces**.
4. The veins, fibrovascular bundles, cut transversely or obliquely.
5. The lower epidermis, much like the upper, but with stomata at intervals.

What is the relation between the position of the **stomata** and the intercellular spaces?

Flower.

Geranium.

I. The General Structure.
    (*a*) Borne on a short stalk (**pedicel**).
    (*b*) Composed of four rows or **whorls** of organs.
1. The external green **calyx**.
2. Inside the calyx the **corolla**, the most conspicuous part of the flower.
3. Inside the corolla the **stamens**.

4. Inside the stamens and forming the middle of the flower, the **pistil**.

II. The Calyx.

Five green-pointed **sepals** attached around the outer edge of the **receptacle** (expanded end of flower-stalk).

Remove two sepals. Sketch.

III. The Corolla.

Consisting of five brightly colored parts, **petals**. Observe that their insertion on the receptacle alternates with that of the sepals.

Remove two or three petals. Sketch.

IV. The Stamens.

Ten in number, each consisting of a stalk-like part, the **filament**, terminated by a small expanded part, the **anther** The filaments are united along part of their length. The anther consists of two lobes or **thecæ**, and a very narrow **connective**.

Sketch a stamen.

Tease out an anther in water and examine with D2.

Numerous **pollen-grains** will be found.

Sketch.

V. The pistil is surrounded by a tube formed by the united filaments of the stamens. Take the stamens off carefully. The lower two thirds of the pistil is stout and green. The upper third is slender and five-cleft at the top (**stigmas**). The **ovary** is deeply lobed. *Sketch.*

Make a transverse section through the ovary.
How many chambers has the ovary? Where are the **ovules** attached? *Sketch.*

VI. *Make a diagrammatic sketch of the flower in longitudinal section, showing the relative position of all its parts.*

Make a diagram of the transverse section of the flower.

In like manner examine, sketch, and make diagrams of the violet, of a papilionaceous flower, and of a flower of the lily or of the amaryllis family, following in each case the order observed for the geranium, and noting the differences.

For description of irregular parts, etc., the student is referred to Gray's "Manual of Botany," under Violaceæ, Leguminosæ, Liliaceæ, Amaryllidaceæ; and to Müller, "The Fertilization of Flowers," under Violarieæ, Leguminosæ, Liliaceæ, and Amaryllideæ.

GERMINATION OF POLLEN-GRAINS.

Mount pollen-grains of narcissus or of daffodil in a 5 per cent. solution of cane-sugar in

water, and of pansy or of violet in a 30 *per cent. solution. Support the cover-glass.*
Examine at once, first with the low and then with the high power. Look for the nuclei. Examine at the end of an hour, and again after another hour.
Sketch different stages.

Seeds.

Examine seeds of the bean, of the pea, of buckwheat, of Indian corn, and of wheat that have been softened in water. Examine also dry seeds of the castor-oil plant.

I. Bean.
 1. Observe on one side the oval spot (**hilum**) to which the stalk (**funiculus**) that fastened the seed to the pod was attached.
 2. Find near the hilum a minute opening, the **micropyle**. If the bean is slightly squeezed a drop of liquid may be pressed out through the opening.
 Cut the seed-coats along the convex edge of the bean, remove them, and examine their inner surface for the internal opening of the micropyle.
 3. Find the **chalaza**—the base of the **nucellus** where the seed-coats blend with each other.

4. Split apart the **cotyledons** and observe the **radicle** and the **plumule** lying between them. *Sketch.*

What is the position of the radicle with reference to the micropyle?

II. **Pea.**

Examine in the same way as the bean.

III. **Castor-oil Bean.**

Carefully take off the hard outer seed-coat and then strip off the thin inner coat. Separate the bean longitudinally into two parts. Examine with a hand-lens.

Find the plumule, the radicle, and the cotyledons. Is there anything else present? What differences are observable between the castor-oil bean and the common bean?

IV. **Buckwheat.**

With a small scalpel take off the outer seed-coat. Observe the thin, light-colored coat beneath. Remove this carefully, and find the parts enclosed by it. How do they compare with the parts within the seed-coats of the bean and pea? And with those of the castor-oil bean?

V. **Indian Corn.**

1. How do the two sides of the grain differ in appearance? *Sketch.*
2. *With a scalpel peel off the seed-coats.*

Observe the yellow **endosperm**. Remove this,

and find the organ of absorption (**scutellum**) enclosing the rest of the embryo and in close contact with the endosperm.

Dissect out the parts enclosed by the scutellum.
Observe :

(a) The radicle directed toward the small end of the grain, and the **root-sheath** covering its free end.

(b) The plumule at the opposite end of the embryo. Its outer leaf is the cotyledon.

(c) The **caulicle**, the attachment between the scutellum and the rest of the embryo.

3. *With a razor make a median longitudinal section through the broad sides of a grain.* Observe on the cut surface :

(a) The tough outer membrane, composed of the united coats of the fruit and the seed.

(b) The endosperm, consisting of starch and other food-materials.

(c) The embryo, with its organ of absorption, the scutellum.

Sketch the section much enlarged.

VI. Wheat.

Examine the seed entire, and then find the parts of which it consists.

Sketch.

SEED-CONTENTS.

I. Pea.

Cut thin sections of the cotyledons. Mount some in dilute glycerine and others in water.

1. In the glycerine preparations observe in each cell:

    (*a*) Large starch-grains.
    (*b*) Small **aleurone-grains**.
    (*c*) Very finely **granulated substance**.

    Observe the intercellular spaces.

    Sketch one or two cells with cell-contents.

2. Compare the glycerine-preparation with the water preparation.

    Add iodine to the glycerine preparation. What is the effect?

II. Bean.

Examine as Pea.

III. Wheat.

Cut thin sections at right angles to the seed-coat, and mount them in dilute glycerine.

Observe:

   (*a*) Just within the seed-coat a layer of rectangular cells containing aleurone-grains.

   (*b*) More centrally situated, cells containing starch. Add iodine. What is the effect?

Sketch a few cells.

IV. Castor-oil Bean.
Gently remove the brittle seed-coat, separate the cotyledons, and make very thin sections of the endosperm.
Mount some in pure glycerine and some in water. Put others into a drop of alcoholic solution of eosin on a slide. After five minutes remove the eosin with filter-paper, add 50 per cent. alcohol and a drop of glycerine, and put on a cover-glass.
1. Study first the glycerine preparation. Observe: large aleurone-grains containing spherical bodies (**globoids**).
2. Compare the water preparation with the glycerine preparation. What difference is observable?
   Add iodine to the glycerine preparation. What is the effect?
3. In the stained preparation look for **crystalloids** in the aleurone-grains.

Sketch.

SEEDLINGS.

I. Dicotyledons.
A. Observe in the earlier stages the way in which the seedling breaks through the ground.
Examine and sketch different stages of each of the following forms:
1. Pea.
2. Bean.

3. Buckwheat.
4. Mustard.
B. Comparison of different kinds of seedlings.
(After examining and sketching the second compare it with the first. In like manner compare the third with the first and second, and the fourth with the first, second, and third.)
1. Compare the roots.
2. Has each a stem supporting the cotyledons (**hypocotyl**)?
3. Compare the cotyledons:
    (a) In regard to form and size.
    (b) What might be inferred as to their function?
4. Compare the cotyledons of the pea and of the bean.
5. Compare the pea and the buckwheat as to the time of appearance of the foliage leaves.
To what may the difference be due?
II. Monocotyledons.
Observe and compare different stages of the following forms:
1. Indian corn.
2. Wheat.
III. Compare the seedlings studied in I. with those in II.
1. Observe the number of cotyledons.
2. Compare their form with that of the foliage leaves.

3. Note the venation of the foliage leaves and their arrangement on the stem.
4. What is the character of the root-system?

ROTATION AND CIRCULATION OF PROTOPLASM.

I. *Mount in water a cluster of young leaves of Nitella. Examine with AA 2 and D 2.*
Observe:
1. The chlorophyll-grains arranged in rows.
2. The neutral zone, free from chlorophyll.
3. The current of protoplasm just below the level of the chlorophyll-grains.

Diagram!

II. *Mount in water one or more hairs from the filament of the stamen of Tradescantia.*
With AA 2 note the rosary-like appearance of the hair. Study with the high power and see the active circulation of currents of protoplasm around and through the central vacuole.

Diagram!

STARCH WITHIN CHLOROPHYLL-GRAINS.

I. *Mount in water some pieces of Nitella leaves.*
Note especially at the cut ends the chlorophyll-granules containing highly refractive bodies—starch-grains.
1. Draw some iodine solution under the cover-glass. What parts are stained?

Diagram!

2. Mount another specimen of Nitella and add

chloral-hydrate iodine. Watch a few minutes for the effect.

KARYOKINESIS.

1. The tip of a young onion-root.*
Find and sketch seven or eight stages.
2. The testis of the lobster.†
Find the principal stages.
Sketch.

---

* Method of preparation: Cut off a quarter of an inch of the tip of young roots of an onion growing in water in a hyacinth-glass; put them into Hermann's fluid in the dark for 48 hours; wash in running water 24 hours; imbed in paraffine and stain on the slide with Haidenhain's iron-hæmatoxylin, leaving the sections in the hæmatoxylin solution overnight.

† Pieces of the testis a quarter of an inch long should be preserved and stained as the onion-root. In making the sections the pieces of testis should be cut *lengthwise.*

FROG.

For an economical use of material this order may be observed:

For the first day's work follow sections I–III inclusive, and in VI the directions for exposing and hardening the central nervous system. This frog should then be put into formic aldehyde solution of 1 to 2 per cent., which is especially good for hardening the nervous system. If the odor of the formic aldehyde is unpleasant, the material may be rinsed in dilute ammonia before being used.

Taking a fresh frog on the second day, work out upon it the circulatory and urino-genital systems, sections IV and V. This frog should be kept in Wickersheimer's fluid, which preserves it in good condition for the study of muscles, etc.

After a few days the brain, etc., of the first frog will have become sufficiently hardened, and the work may proceed in regular sequence. The directions after VI may be worked out upon either frog, as the material serves.

For this work on the frog it is sufficient to use a water solution containing from 1 to 2 per cent. of formic aldehyde. "Formalin" is properly a commercial name for a 40 per cent. solution of formic aldehyde in water, but is often used indiscriminately for solutions of various strengths.

I. External Characters.
1. The division of the frog into head, trunk, and limbs.
2. The skin: moist and smooth. Note its looseness.
Make short slits in several places and inflate the lymph-sacs.
3. The head: its shape.
    (a) The eyes: each with two eyelids (are they movable?).
    (b) The **tympanic membranes**; modified parts of the skin covering the ears.
    (c) The nostrils or **anterior nares**.
4. The limbs.
    (a) The anterior limb is divided into three regions: **brachium, antibrachium,** and **manus.** The manus has four **digits.** In the breeding season the first digit of the male bears a swollen cushion.
    (b) The posterior limb is divided into three regions: **femur, crus,** and **pes.** The pes has five digits connected by a web.
5. Identify all openings into the body.
6. Look at a living frog in an aquarium-jar with a little water to see its position, its eyes, etc., and its mode of breathing.

Make an outline sketch of the dorsal aspect of the frog.

II. **The Buccal Cavity.**

Open the mouth and note:
1. Its wide cavity; the posterior part, the pharynx, is continuous with the œsophagus.
2. The teeth:
    (a) On the upper jaw.
    (b) On the roof of the mouth.
3. The **posterior nares.**
    To find them, pass bristles into the anterior nares and note the points of entrance into the mouth-cavity.
4. The **Eustachian recesses:** a pair of lateral prolongations of the mouth-cavity. Pass a bristle through the tympanum from the outside and note the place of its entrance into the mouth-cavity.
5. The tongue. Note its shape and its attachment. Turn it forward to see:
6. The **glottis:** a longitudinal slit in the floor of the mouth.

Pass a probe into it.

III. The Abdominal Viscera.

Lay the frog upon its back in a pan containing wax, and fasten it down with pins through the limbs. Cut through the skin along the median ventral line, and make a transverse cut at each end of the first. Turn the flaps of skin outward and pin them down.

Observe:

1. On the skin near the shoulder the **musculocutaneous** vein.
2. The body-wall, formed of muscles.
3. In the median ventral line the **anterior abdominal** vein.

Put a double ligature on this by making a slit on each side of the vein and tying two threads around it near together. Raise the body-wall with forceps, and with scissors carefully continue the cut on the left side of the vein from pelvis to sternum. Make a transverse cut between the ligatures and extending across the ventral aspect of the body.

Note the anterior abdominal vein passing dorsally into the liver.

With scissors cut in the median line through the sternum and other superficial parts, taking care not to injure the parts beneath. Turn each half outward and pin firmly.

4. Sketch the organs exposed to view and identify them. (*Remove the ovaries if they conceal the other organs.*)
    (*a*) The heart.
    (*b*) The liver: reddish brown and bilobed.
    (*c*) The lungs.

(d) The stomach: lying partly beneath the left lobe of the liver.

(e) The small intestine: a light-colored slightly convoluted tube. It passes posteriorly into the large intestine. The terminal part of the large intestine is the cloaca. Note the mesentery by which the intestine is attached to the dorsal body-wall.

(f) The bladder: a thin-walled bilobed sac in the posterior ventral part of the body-cavity.

Insert a blowpipe into the glottis and inflate.

The lungs, if hidden before, will now come into view.

Pass a probe into the mouth and through the œsophagus into the stomach. Turn the liver forward to see the œsophagus and stomach. Uncoil the intestine and identify its parts. Find:

(g) The pancreas, in the loop between the stomach and duodenum.

(h) The spleen, in the mesentery near the beginning of the large intestine.

(i) The two lobes of the liver. Observe that the larger lobe is subdivided into two. Note the position of the gall-bladder. Make a slight slit in the duodenum and try by pressing upon the gall-bladder to make the bile pass to the duodenum. Look now for the bile-duct.

(j) The fat bodies: long slender yellow masses on each side, attached to the dorsal wall of the body-cavity at about the level of the posterior border of the liver. In different specimens they vary much in size.

Identify:

(k) The reproductive organs.

IV. The Circulatory Organs.

Open the pericardial cavity, using great care not to injure the anterior abdominal vein which passes dorsally in the posterior wall of the pericardium.

1. Examine and sketch the heart, showing its parts:

    (a) The auricles: dark colored and with thin walls.

    (b) The ventricle: of paler color and conical.

    (c) The **truncus arteriosus**: arising from the anterior side of the ventricle, passing obliquely forward and dividing into two large branches.

    Turn the apex of the heart forward and note on its dorsal side:

    (d) A darker triangular-shaped region, the **sinus venosus**. It is closely related to the auricles, and receives the great veins.

2. Observe the contractions of the heart. Note the order of sequence of the contraction in the four parts of the heart. The heart of lower vertebrates often continues to beat after the death of the animal.
3. The veins.
   A. The Anterior Abdominal Vein and its Tributaries.
   (a) Trace the course of the anterior abdominal vein into the liver. Just before its entrance it divides into two branches, one going to each main lobe of the liver.
   (b) Trace it posteriorly. It is formed by the union of two **pelvic** veins which are ventral branches of the **femoral** veins.
   (c) Trace the pelvic vein of one side back to the femoral from which it arises.
   (d) Follow the femoral vein posteriorly, to see from what region it is bringing blood.
   B. The Renal Portal System.
   (a) The dorsal branch of the femoral vein is the **renal portal** vein. It receives the **sciatic** vein and small veins from the body-wall, etc. The renal portal vein passes forward to the outer side of the kidney and enters that organ by means of numerous branches.
   C. The Hepatic Portal System.

(a) Raise the liver and find the vein that enters its left lobe, the **hepatic portal vein**. Trace it posteriorly and find the veins that unite to form it:
 (aa) The **gastric** vein: from the stomach.
 (bb) The intestinal veins.
 (cc) The **splenic** vein.

The hepatic portal vein just before entering the liver gives off a branch that anastomoses with the anterior abdominal vein.

D. Veins Opening into the Sinus Venosus.

(a) The **inferior vena cava**: a median vein which opens into the posterior part of the sinus venosus. It brings blood to the heart from the liver and kidneys, etc., and from the hind limbs. It is made up by:
 (aa) The **renal** veins: from the kidneys.
 (bb) The **genital** veins.
 (cc) The right and left **hepatic** veins from the liver. These open into the vena cava inferior close to the sinus venosus.

(b) The right **superior vena cava**. It brings back blood from the right side of the head and body and from the right fore-limb. It is formed by:
 (aa) The **external jugular** vein: formed by:

(*aaa*) The **lingual** vein: from the tongue.

(*bbb*) The **inferior maxillary** vein: from the lower jaw.

(*bb*) The **innominate** vein: formed by

(*aaa*) The **internal jugular** vein: from the brain, etc.

(*bbb*) The **subscapular** vein: from the muscles of the shoulder.

(*cc*) The **subclavian** vein: formed by

(*aaa*) The **brachial** vein: from the forelimb.

(*bbb*) The musculo-cutaneous vein from the skin, etc.

(*c*) The left **superior cava**: similar to the right.

E. Vein Opening into the Left Auricle.

(*a*) The **pulmonary** vein: formed by the union of the right and left pulmonary veins. Each pulmonary vein runs along the inner side of its lung.

Diagram of the veins!

4. The Arteries.

Distend the œsophagus by inserting the rubber top of a pipette or a roll of paper, and thus stretch the arteries.

Note the division of the truncus arteriosus into two branches, and that each of these is subdivided into three aortic arches:

(*a*) The anterior or **carotid arch**. Its chief branches are :
(*aa*) The **lingual** artery : to the tongue.
(*bb*) The **carotid** artery : to the brain, etc.
(*b*) The middle or **aortic arch**. It runs around the throat to the vertebral column and unites with its fellow to form the **dorsal aorta**. It gives off on each side at the level of the arm :
(*aa*) The **subclavian** artery : to shoulder and fore-limb.
(*bb*) The **vertebral** artery : to the vertebral column and back, etc.
(*c*) The third, posterior, or **pulmonary arch** : to the lung on each side. On the way it gives off :
(*aa*) The **cutaneous** artery : to the skin, etc.
(*d*) The dorsal aorta and its branches.
Posterior to the union of the two aortic arches the dorsal aorta lies in the median line just ventral to the vertebral column. It gives off the following branches :
(*aa*) The **cœliaco-mesenteric** artery : a median artery to the stomach and intestine. Its branches are :
(*aaa*) The **cœliac** artery : to the stomach and to the liver.

(*bbb*) The **mesenteric** artery: to the intestine and to the spleen.

(*bb*) The **renal** arteries: to the kidneys.

(*cc* The **genital** arteries: to the reproductive organs.

(*dd*) The **inferior mesenteric** artery: supplying the base of the large intestine.

(*ee*) The **common iliac** arteries: formed by the division of the dorsal aorta. Each continues down the leg as the **sciatic** artery after having given off:

(*aaa*) The **hypogastric** artery: to the bladder.

Diagram of the Arterial System!

V. The Urino-genital Organs.

Remove the alimentary canal between the base of the œsophagus and the posterior third of the large intestine.

Posterior to the fat bodies are the reproductive organs.

In the female:

1. The ovaries: lobed organs, varying much in size with the season of the year.
2. The oviducts: convoluted tubes, one on each side. The anterior end is funnel-like, and the tube passes posteriorly to open into the cloaca.

3. The kidneys: elongated red masses close to the vertebral column.
4. The **adrenal bodies**: a band of yellow tissue on the ventral side of each kidney.
5. The **ureters**: two whitish tubes passing from the outer edge of the kidney to the cloaca.

In the male:
1. The testes: a pair of yellowish rounded bodies.
2. The genital ducts (**vasa efferentia**) placing each testis in communication with the kidney of the same side. In the male the ureter serves also as a genital duct.
3. The kidneys, as in the female (see above).
4. The adrenal bodies, as in the female (see above).
5. The ureters, as in the female (see above).

Diagram of the Reproductive Organs!

VI. The Nervous System.

A. The Central Nervous System.

Cut the skin along the median dorsal line and turn it back. Remove the muscles from the vertebræ. Open the neural canal by cutting the membrane between the skull and the first vertebra. Remove the roof of the brain-cavity gradually, bit by bit, with small strong forceps or with small strong scissors. In the same way remove the neural arches of the vertebræ.

Note the delicate pigmented membranes that cover the brain and the spinal cord. The outer, the **dura mater**, is often more or less torn in removing the bone. The inner, the **pia mater**, lies very close upon the nervous tissue. It is usually more pigmented than the dura mater.

Before the brain can be removed for further investigation it must be hardened by exposure to formic aldehyde.

To remove the brain :

Cut through the olfactory tracts (or nerves) close to the nasal pits ; lift the anterior end of the brain gently, cut through the cranial nerves close to the skull and through the spinal nerves. Take out the brain and spinal cord and put into a small dish of water (or dilute alcohol).

Sketch the nervous system thus exposed :

1. The brain : the dorsal aspect (from before backward):
    (a) The **olfactory lobes**, each passing anteriorly into a cylindrical portion : the **olfactory tract**.
    (b) The **cerebral hemispheres**, separated from each other by a deep cleft. They are marked off from the olfactory lobes by a slight transverse depression.
    (c) The **thalamencephalon**: immediately behind

the cerebral hemispheres. In the middle line between the diverging ends of the cerebral hemispheres the **pineal gland** is borne.

(d) The **optic lobes**: two conspicuous rounded bodies, one on each side.

(e) The **cerebellum**: a narrow ridge posterior to the optic lobes.

(f) The **medulla oblongata**: the region posterior to the cerebellum, broadest at the anterior end and passing gradually into the spinal cord. Its dorsal wall is a thin, highly vascular membrane, the **choroid plexus** of the fourth ventricle. Beneath the choroid plexus is the cavity of the **fourth ventricle**.

2. The Spinal Cord: continued posteriorly from the medulla oblongata. It is flattened dorso-ventrally. Near the region of the sixth vertebra it tapers rapidly to a slender thread, the **filum terminale**. The filum terminale and the proximal ends of the lumbar nerves form the **cauda equina**. Note in the median line a slight groove not well marked throughout, the **dorsal fissure**.

Turn the brain and spinal cord ventral side uppermost, and sketch. Note:

3. The Brain:
   (a) The olfactory lobes and nerves.

(*b*) The cerebral hemispheres.
(*c*) The **optic chiasma**: ventral to the posterior end of the hemispheres. Trace the bundles of fibres as far as possible.
(*d*) The **tuber cinereum**: a shield-shaped body posterior to the optic chiasma. On its posterior border attached by a slender stalk:
(*e*) The **pituitary body.**
(*f*) On each side of the pituitary body the **crura cerebri**: two columns of fibres connecting the cord and the medulla with the anterior part of the brain.
4. The Spinal Cord, continued posteriorly from the medulla. Note the median (ventral) fissure throughout its length.
5. The Cavities of the Brain.

With a razor make a horizontal cut midway between the dorsal and the ventral sides to show:

(*a*) In the cerebral hemisphere on each side the **lateral ventricle.**
(*b*) Connecting the two lateral ventricles the **foramen of Monro.**
(*c*) In the optic lobes the **ventricles of the optic lobes.**
(*d*) Connecting (*a*) and (*c*) the **third ventricle.**
(*e*) In the medulla the **fourth ventricle.**
B. The Sympathetic Nervous System.

Lay the frog ventral side up and carefully remove the digestive, excretory, and reproductive organs.

Observe:

1. Along the vertebral column on each side and close to the dorsal aorta a slender cord.

    With great care raise it slightly and note:

    The yellowish enlargements along its length, the **sympathetic ganglia.**

    Count the ganglia, being careful not to injure them.

C. The Peripheral Nervous System.

1. Spinal Nerves: passing out from between the vertebræ. Count those of one side.

    Gently raise the sympathetic cord, and note the branch from each spinal nerve to a sympathetic ganglion.

    Note the grouping of the spinal nerves into plexuses:

    (*a*) The **sciatic plexus**: formed from the 7th, 8th, and 9th spinal nerves. From the plexus the **sciatic nerve** is given off. Trace this into the leg.

    (*b*) The **brachial** plexus: formed from the 2d and 3d spinal nerves. It gives rise to the **brachial nerve.** Trace this nerve into the arm.

    Between these two plexuses the 4th, 5th, and 6th spinal nerves pass out separately as small nerves to the muscles and skin of the body-wall.

Diagram of the sympathetic ganglia and their principal nerves, and of the spinal nerves of one side.

VII. The Eye.

With fine scissors make an incision in the skin above the eyeball. Continue this cut around the edge of the socket. Lifting the eyeball by the edge of the skin, cut it free from the muscles and nerve beneath. Take out both eyes and harden them for two or three days in formic aldehyde solution.

Divide one eye into right and left halves, and one into proximal and distal halves.

Observe:
1. The **sclera**.
2. The **cornea**.
3. The **choroid coat**.
4. The **retina**.
5. The **lens**.
6. The **vitreous humor**.

VIII. Muscles.

Cut through the skin around the junction of the leg with the body, and strip the skin from the leg.

Bend the leg, and try to distinguish the flexors and extensors. Dissect the leg to show the **sartorius** and **gastrocnemius** muscles. (See diagram.)*

---

* Enlarged from Figs. 80 and 81, Ecker's "Anatomy of the Frog," translated by G. Haslam. Clarendon Press, 1889.

How does a muscle begin and how does it end? Compare the middle with the end.

IX. Bones.

Examine the articulation of the femur with the pelvis. Observe the end of the bone, cartilage, etc.

Clean up the knee-joint, the pelvis, one or two vertebræ, and the skull, and see their parts.

FISH.

I. External Characters.
  1. Make an outline sketch showing the fins, orifices, etc.
  2. Find the **operculum**. Lift it up and find :
  3. The **gills**. How many are there? How many slits between the gills (**gill-clefts**)?

II. Abdominal Viscera.
  Cut through the body-wall on both sides along the lines indicated in the diagram.* Remove the portion of body-wall thus cut out.
  Observe :
  1. The liver, a conspicuous brownish lobed organ. Lift it up to see :
  2. The gall-bladder.
  3. The hepatic veins, passing anteriorly from the liver.
  4. The stomach, partly concealed by the liver.
  5. The intestine, passing posteriorly from the stomach with one or two turns to the anus.

---

* An outline-diagram of the fish with lines to mark places for cutting :
  1. Antero-posteriorly : parallel and slightly ventral to the lateral line.
  2. Dorso-ventrally from the median ventral line to meet the first cut :
     (a) An inch in front of the anus.
     (b) Obliquely just posterior to the pectoral and pelvic girdles.

6. The pyloric appendages, finger-like organs opening into the alimentary canal at the junction of stomach and intestine.
7. The spleen, a dark red body lying in one of the loops of the intestine.
8. The reproductive organs.
9. The air-bladder.

Diagram !

Cut through the intestine an inch from the anus and through the hepatic veins close to the liver, and remove all the organs thus far seen except the air-bladder.

Cut through the air-bladder (noting the **rete mirabile** *on its ventral wall) to see:*

10. The kidneys, elongated red organs pressed close to the dorsal wall of the abdominal cavity, one on each side of the vertebral column. Put a bristle into the outer opening of the ureter and trace it to the kidneys.

Find:

11. The genital duct.

Diagram of the kidneys and reproductive organs.

III. Circulatory System.

Dissect away the left half of the pelvic and pectoral girdles from below upward, not injuring the veins.

Observe:

1. The pericardial cavity and the septum separating it from the abdominal cavity.
2. The heart.
    (a) The sinus.
    (b) The auricle.
    (c) The ventricle.
    (d) The bulbus (conus).
Diagram
3. Find the veins coming to the heart:
    (a) The hepatic veins, passing through the septum from the abdominal cavity and opening into the sinus.
    (b) The pre-caval veins (**ducts of Cuvier**), one on each side formed by the union of the jugular vein from the anterior and of the cardinal vein from the posterior region, passing ventrally along the septum and opening into the sinus.
4. Trace the ventral aorta from the bulbus and find its branches.
5. Find the dorsal aorta (dorsal to the air-bladder) and trace it forward into the cephalic circle.

Diagram of the veins and arteries.

IV. Exoskeleton.

Take off several scales from different parts of the body. Examine with a low power.

Sketch.

PIGEON.

I. External Characters.
  1. Note the shape of the head, the neck, the trunk, and the tail.
  2. Note the arrangement of feathers in the body, the wing, and the tail. Pull out some of each of the different kinds and keep them for examination.
  3. Sketch the head. How many openings? Compare with those of the frog.
  4. Observe the different parts of the wing and of the leg. Find the hand and the foot. Note the attachment of the feathers. What forms the covering of the foot?
  5. Pluck.
  (*This can be done more easily if the skin of the bird is first thoroughly wet with very hot water. Great care should be taken when pulling out feathers from the neck not to tear the skin of that region.*)

II. The Abdominal Viscera.
  Make a median longitudinal cut through the skin on the ventral surface of the body and neck, being careful not to cut deeper.
  Turn back the skin, and keeping it moist, note:
  1. The œsophagus and crop.
  2. The trachea.

PIGEON.

3. The blood-vessels (be careful not to injure them).

Slit the trachea transversely in the middle neck-region and ligature the neck through the slit and under (i.e., not including) the skin. Cut off the head above the ligature. Open the brain-cavity by shaving off the bone, exposing cerebrum and cerebellum, and put the head at once into formic aldehyde solution (as for frog's brain).

Make a transverse cut through the body-wall of the abdomen posterior to the pectoral muscles.

4. Note the air-sacs. How many are visible?
5. Observe the intestine and the fat in the mesentery.
6. Find the **falciform ligament**, a thin membrane with a large vein, in the median vertical plane and attached to the middle line of the **sternum** beneath the **keel**.
7. *Insert a tube into the trachea and blow up the structures connected with it.* (Compare II. 4., above.)

Holding the sternum with the right hand press gently with the left thumb upon the part of the crop that lies against the pectoral muscles and thus separate the crop from the anterior border of the sternum.

Note beneath the crop:

8. A bilobed sac in the angle formed by the furcula, the **interclavicular air-sac**.
9. Dorsal to each lobe of the interclavicular air-sac is one of the paired **prebronchial air-sacs**.

*Cut through the left **pectoralis major** muscle close to the keel of the sternum along its whole length, half an inch deep at the anterior end and less deep as the posterior end is approached. From the anterior end continue the cut laterally and dorsally to separate the pectoralis major from the furcula. From the posterior end of the keel continue the cut laterally to separate the muscle from the body of the sternum. With the handle of a scalpel gently separate the pectoralis major from the **pectoralis minor** beneath.*

Note:
10. The **pectoral** blood-vessels.
11. The **axillary air-sac**.
12. Note the œsophagus passing posteriorly from the crop.

*Cut carefully through the remaining muscle (**pectoralis secundus**) close to the keel of the sternum. With strong scissors cut through the body of the sternum close to the keel; cut through the furcula and cut away the articulation of the coracoid with the sternum.*

PIGEON. 119

Reflect the two sides of the body-wall and note the viscera now exposed to view:
13. The heart in the pericardium.
14. The liver. How many lobes has it? Which one is larger?
15. The duodenum, a U-shaped loop of the intestine.
16. The pancreas, lying between the limbs of the loop. Look for its ducts.
17. The gizzard.
18. The loops of the small intestine.
19. The **epigastric** vein, carrying blood from the omentum and passing up in the **falciform** ligament to the anterior border of the liver.

(Note: 20 and 21 should be left until after III.)

20. Unravel the intestine and observe:

Œsophagus.   Intestine.
Crop.   Rectum and **rectal cœca.**
Proventriculus. Liver.
Gizzard.   Pancreas.
Spleen.

Diagram of the abdominal viscera, ducts, etc.!

21. Cut open the gizzard and observe its structure.

III. The Circulatory System.
   The Heart.
   Open the pericardium and observe the heart:

Auricles.
Ventricles.
Diagram!
The Veins.

A. The **anterior venæ cavæ**, one on each side opening into the right auricle; formed by the union of:
  1. The **jugular** vein, bringing back blood from the head and neck, receiving:
  (a) The **vertebral** vein.
  2. The **brachial** vein, bringing blood from the wing.
  3. The **pectoral** vein, bringing blood from the chest muscles.

B. The **posterior vena cava**, an unpaired vein, opening into the right auricle; formed by:
  1. The two **iliac** veins. Each iliac vein begins near the anterior end of the kidney by the union of:
    (a) The **femoral** vein, bringing blood from the leg.
    (b) The **hypogastric** vein, passing through the kidney and formed in the posterior abdominal region by the union of:
      (aa) The **caudal** vein, an unpaired vein from the tail.
      (bb) The **internal iliac** vein from the pelvis.
      (cc) The **posterior mesenteric** vein (**coccygeo**

mesenteric), an unpaired vein from the lower part of the intestine.

    (*dd*) The **sciatic** vein, from the leg; opening into the hypogastric vein near the junction of the middle and posterior lobes of the kidney.

  (*c*) The **renal** veins, bringing blood from the kidneys, a larger vein from the posterior lobes, and a smaller vein from the anterior lobe.

2. The **hepatic** veins from the liver; they join the post cava near its entrance into the right auricle.

C. The **hepatic portal system.**
  1. The **gastric** veins, two small veins from the left side of the gizzard to the left lobe of the liver.
  2. The **portal** vein, sending a branch into the right and a branch into the left lobe of the liver, formed by:
     (*a*) The **gastro-duodenal** vein from the right side of the gizzard, duodenum, pancreas, etc.
     (*b*) The **anterior mesenteric** vein, from most of the small intestine.
     (*c*) The **posterior mesenteric** vein from the large intestine, etc.

  The portal vein receives blood through a small vein from the spleen.

D. The **pulmonary** veins, short veins from the lungs which unite and open into the left auricle.

Diagram of the veins!

The Arteries.

A. One aortic arch, the **aorta**, which passes to the right. It gives off:
  1. The **innominate** arteries, one on each side, giving rise to:
     (a) The **common carotid** artery, to the head.
     (b) The **subclavian** artery, which divides to form:
       (aa) The **brachial** artery, to the wing.
       (bb) The **pectoral** artery, to the muscles of the chest.
  2. The **dorsal aorta**, which passes posteriorly dorsal to the heart.
     It gives off:
     (a) The **cœliac** artery, unpaired, to the proventriculus, gizzard, spleen, pancreas, duodenum, etc.
     (b) The **anterior mesenteric** artery, unpaired, to the rest of the small intestine.
     (c) The **anterior renal** arteries, paired.
     (d) The **femoral** arteries, paired.
     (e) The **sciatic** arteries, paired. The sciatic artery of each side gives off:
       (aa) The **middle** and **posterior renal** arteries.
     (f) The **posterior mesenteric** artery, unpaired, to the large intestine.

(g) The **internal iliacs**, paired, to the pelvis.

(h) The **caudal** artery, the posterior continuation of the dorsal aorta.

B. The pulmonary artries, to the lungs.

Diagram of the arteries!

N. B.—Return here to II. 17 and 18.

IV. The Excretory and Reproductive Organs.

Carefully remove the digestive organs, cutting through the intestine about an inch from its posterior end.

1. The ovaries and oviducts, or testes and vasa deferentia.
2. The kidneys, adrenal bodies, ureters.
3. The cloaca.

Slit open the rectum on the ventral side and continue the cut into the cloaca.

(a) Note the number of chambers into which the cloaca is divided.

(b) Find the openings of the ureters on the dorsal side of the cloaca in the urino-genital pouch. Also, the openings of the genital ducts.

Diagram of excretory and reproductive organs!

V. The Nervous System.

A. The Sympathetic Nervous System.

1. A delicate cord running on each side close to the vertebral column, with ganglia at intervals.

Diagram!
B. The Peripheral Nervous System.
1. On the dorsal wall of the body-cavity note: the **thoracic spinal nerves** coming from between the vertebræ and passing out between the ribs.
2. The **brachial plexus,** a network of nerves at the base of the neck. Trace the distribution of the chief nerves given off from it.
3. The **lumbar plexus** in the lumbar region. Two nerves are given off from it; what do they innervate?
4. The **sciatic plexus.** The sciatic nerve is given off; what is its course?

Diagram!
C. The Brain.
1. Externals.
As the brain lies in the cranial cavity, note:
(*a*) The cerebral hemispheres.
(*b*) The olfactory nerves.
(*c*) The optic lobes.
(*d*) The cerebellum.
(*e*) The medulla oblongata.

Diagram!
Remove the dorsal arches of the first two vertebræ, cut through the spinal cord, and turn the brain forward, carefully cutting all the nerves.
On the ventral side note:

(*f*) The optic chiasma.
(*g*) The infundibulum.
(*h*) The pituitary body.
Diagram !
2. Section.
(*a*) Gently press apart the hemispheres ; note that they are not connected.
Make a shallow incision in the inner wall of one of the hemispheres near its postero-dorsal corner so as to open its cavity. Carefully cut away enough of the inner end of the posterior wall to see the whole of the cavity.
(*b*) Note the cavity of the cerebral hemisphere : the lateral ventricle. Its floor is formed by the **corpus striatum.**
Remove the cerebellum by cutting through its peduncles.
(*c*) Note :
   (*aa*) The two optic lobes connected by the optic commissure.
   (*bb*) Anterior to the optic lobes the paired optic thalami.
   (*cc*) Between the optic thalami the third ventricle.
   (*dd*) Anterior to the optic thalami is the anterior commissure, and posterior to them is the posterior commissure.
   (*ee*) Posterior to the optic commissure is the fourth ventricle.

(ff) At each side of the fourth ventricle are the peduncles of the cerebellum.

(d) Cut off the dorsal wall of one optic lobe and see its ventricle.

(e) Make a median (vertical) longitudinal section of the brain and see the continuity of the third and fourth ventricles.

Diagram!

VI. The eye.

Divide one into two lateral halves, and the other into anterior and posterior halves.

Observe:
1. The two humors.
2. The sclerotic coat.
3. The choroid coat.
4. The retina.
5. The ora serrata.
6. The entrance of the optic nerve.
7. The pecten.
8. The ciliary processes.
9. The lens.
10. The iris.
11. The pupil.

VII. The Feather.

In a tail feather, note:
1. The stem (**scapus**), consisting of:
 (a) The quill (**calamus**), the hollow proximal part.

(*b*) The **shaft** (**rachis**), the solid distal part.
2. The **vane** (**vexillum**), made up of **barbs**.
3. The **inferior** and the **superior umbilicus**.

Cut off a piece of the vexillum, soak it for a few minutes in a small dish of alcohol to remove the air, and examine in glycerine under a high power.

Note the barbs, **barbules**, and **barbicels**. What is the relation between the barbules and the barbicels?

Sketch, noting which is the proximal and which the distal side?

RABBIT.

I. External Characters.
   1. The body consisting of:
      (*a*) The head with:
         (*aa*) The mouth; note the character of the lips.
         (*bb*) The nostrils; note their relation to the upper lip.
         (*cc*) The **vibrissæ**, or whiskers.
         (*dd*) The ears, with elongated **pinnæ**.
      (*b*) The neck,—short and thick.
      (*c*) The trunk, made up of:
         (*aa*) The thorax, the anterior part, protected by bony structures.
         (*bb*) The abdomen, the posterior part, with soft walls. Its openings are:
            (*aaa*) The urino-genital opening in the median line, lying near:
            (*bbb*) The anus, also in the median line, posterior to the urino-genital opening.
         Note in the female the **mammæ** or teats, four or five pairs on each side, near the ventral median line.
         In the male, slightly anterior to and at the sides of the reproductive opening, the **scrotal sacs**.
      (*d*) The tail.
   2. The fore and hind limbs.

Each divided into three chief regions.
Count the number of digits.

II. Abdominal Viscera.

Spread the animal out dorsal side down with its head hanging over the edge of the pan, and fasten it down by the legs.

Make a shallow cut along the median ventral line and skin the ventral and lateral regions of the abdomen and thorax. (The skinning to be completed later.) Open the abdominal cavity by a cut in the median ventral line.

1. What forms the anterior wall of the abdominal cavity?
2. Identify the organs exposed to view; e.g., the stomach, the liver, the intestine, etc.

Pass a single ligature around the œsophagus near the cardiac end of the stomach, and double ligatures around the rectum about two inches from the anus. Find the portal vein, and double-ligature it at a little distance from the liver. (So that later the intestine may be taken out.)

3. Examine the intestine.
   (a) In place, finding:
      (aa) The pancreas and duct.
         The pancreas consists of a number of scattered masses lying in the mesentery of the duodenum. Its duct opens into the ascending loop of the duodenum, and is formed

by the union of numerous smaller ducts from the different lobes of the pancreas.

(*bb*) The gall-bladder and the **common bile-duct.** The gall-bladder lies posterior to one of the lobes of the right side of the liver. Its duct unites with a duct from each lobe of the liver to form the common bile-duct, which opens into the duodenum close to the pylorus.

(*b*) Remove the intestine (for directions see the next following italicized paragraph) and unravel and identify its different parts. Diagrams! In taking out the intestine note the attachment of the mesentery to the intestine and to the body-wall. Compare its extent in one of these regions with its extent in the other.

To remove the intestine cut above the ligature on the œsophagus and between the double ligatures. Ligature any large vein that it is needful to cut through. Be careful not to injure the reproductive organs.

Remove the eyes, carefully, with the muscles and put them into formic aldehyde solution (1 to 2 per cent.).

Skin the head, ligature the vessels of the neck,

and cut off the head. Open the skull to expose the cerebrum and most of the cerebellum. Put the head into formic aldehyde solution (1 to 2 per cent.).

Finish skinning the body (leaving the skin intact in the region of the anus and the scrotal sacs).

III. The Thoracic Cavity.

Open the thorax without injuring the diaphragm by a slit on each side of the sternum. Remove the sternum.

1. Examine the structure of the diaphragm (muscle, tendon).
2. Insert a blowpipe into the trachea and blow up the lungs.

IV. The Circulatory System.

Open and cut away the pericardium. Dissect off the fat and the connective tissue around the heart and the vessels near to it.

In young animals the fat-like **thymus** gland is very conspicuous in the anterior end of the thorax, and sometimes it lies over the anterior end of the heart.

Find the parts of the heart:

The right and left auricles.

The right and left ventricles. (If no division-line can be seen, try to distinguish the two ventricles by feeling the thickness of their walls.)

A. The Veins.
   Leave those in the posterior abdominal region until after the reproductive organs, etc., have been examined.
I. The Venæ Cavæ.
   A diagram should be begun at an early stage in the dissection of the veins, etc., and added to as each new structure is seen.
   The anterior venæ cavæ.
   Opening into the right auricle the **right anterior vena cava**, a large vein bringing blood from the right arm and the right side of the head, etc. It receives near the heart (best seen by turning the heart over to the left):
   1. The **azygos** vein which comes from the median dorsal thoracic region, bringing blood from the posterior intercostal spaces of both right and left sides of the body.
   2. Near the entrance of the azygos a smaller vein, the **right intercostal** vein, bringing blood from the right anterior intercostal region.
   3. Anterior to the intercostal vein a **vertebral** vein from the brain and the spinal cord.
   4. A small vein from the inner side of the ventral wall of the thorax, the **internal mammary** vein. Near the entrance of the internal mammary vein the anterior vena cava is formed by the union of:

5. The **subclavian** vein, a large vein from the arm and shoulder, with:
6. The **external jugular** vein, from the superficial parts of the head. The external jugular vein near its union with the subclavian vein receives:

(*a*) The **internal jugular** vein from the brain.

The **left anterior vena cava**, like the right anterior vena cava, except that it receives a small **coronary** vein from the walls of the heart, and does not receive an azygos vein.

The **posterior vena cava**, a large vein that empties into the right auricle. It receives near the heart:

1. The **hepatic** veins from the lobes of the liver.
2. Veins from the body-wall, one on each side, entering the posterior cava close to or in connection with:
3. The **renal** veins from the kidneys, one on each side.
4. The **spermatic** (or **ovarian**) veins from the reproductive organs.

The veins posterior to the spermatic veins should be left until after the examination of the reproductive organs, etc.

5. The **ilio-lumbar** veins from the posterior body-wall. The left ilio-lumbar vein in some rabbits does not enter the posterior cava at the same level with its fellow of the op-

posite side, but runs anteriorly to about the level of the left renal vein. In this case the left spermatic vein empties into the ilio-lumbar vein.
6. The **external iliac** veins, bringing blood from the legs, from the bladder, and in the female from the uterus.
7. The **internal iliac** veins, which by their union form the posterior vena cava. They bring blood from the postero-dorsal part of the thighs.

II. The Portal System.
The **portal** vein enters the liver by a branch to each lobe. (It is formed by veins from the stomach, from the different regions of the intestine, and from the spleen.)

III. The Pulmonary Veins.
(These are best seen when taking out the heart and lungs, IV. C.) Two veins from each lung which enter the dorsal side of the left auricle.

B. The Arteries.
I. The **aortic arch**, arising from the left ventricle and curving to the left to form in the median dorsal line the **dorsal aorta**.
The aorta gives off the following arteries:
1. Near the beginning of the arch, a vessel to the right, the **innominate** artery, which soon divides to form:

(a) The right subclavian artery. It divides into:
: (aa) The **vertebral** artery which passes dorsally and toward the median line.
: (bb) The **internal mammary** artery to the inner wall of the thorax. It sends off five branches to the five anterior intercostal spaces.
: (cc) The **brachial** artery to the arm.
(b) The **right common carotid** artery.
2. The **left common carotid** artery, on the left of the median line. (Not infrequently united with 1.)
3. The **left subclavian** artery.
  Its branches like those of the right side.
4. The **thoracic** arteries; seven pairs to the seven posterior intercostal spaces.
5. A little posterior to the diaphragm, the large unpaired **cœliac** artery to the liver, the stomach and the spleen.
6. The anterior **mesenteric** artery to the small intestine, the pancreas, the cœcum and the colon. It arises near the cœliac artery.
7. The **renal** arteries, paired, to the kidneys.
8. The **spermatic** or **ovarian** arteries to the reproductive organs.
  (The arteries posterior to the spermatic, etc., arteries should be left until after the

examination of the reproductive organs, etc.)

9. The **posterior mesenteric** artery, unpaired, to the rectum.
10. The **sacral** artery from the dorsal side of the dorsal aorta, unpaired, to the tail.
11. The **iliac** arteries, the two branches into which the dorsal aorta divides at its posterior end. Each of these gives off:
    (a) The **ilio-lumbar** artery to the body-wall, and divides into:
    (b) The **internal iliac** artery, to the pelvis.
    (c) The **external iliac** artery, which gives off an artery to the bladder and becomes the **femoral** artery to the hind leg.

II. The Pulmonary Artery.

(These are best seen when taking out the heart and lungs. IV. C.).

From the ventral side of the right ventricle a vessel passes anteriorly, and, curving around the dorsal side of the heart, divides into right and left branches to the right and left lungs.

Diagrams!

C. The heart. Find its parts again; then ligature the posterior vena cava near the heart and take out the heart.

To remove the heart cut through all the veins and arteries (leaving the ends attached to the heart

as long as possible) except those connected with the lungs. Take out heart and lungs and put them into a dish of water.
1. Identify again the vessels entering and leaving the heart.

Cut away the outer walls of both auricles and wash out the blood.
2. See the septum with the **fossa ovalis**, and the entering veins.

Make with fine scissors a cut through the ventral wall of the pulmonary artery and continue it through the wall of the right ventricle.

Note :
3. The auriculo-ventricular (**tricuspid**) valve.
4. The ventricular septum.
5. The **semilunar** valves at the base of the pulmonary artery.
6. Cut away the outer wall of the left ventricle and compare its thickness with that of the right ventricle.

Note in the left ventricle :
7. The auriculo-ventricular (**mitral**) valve.
8. The **chordæ tendineæ**.
9. The **papillary** muscles.
10. Compare 7, 8, and 9 with the corresponding structures in the right ventricle.

Note :
11. The semilunar valves at the beginning of the aortic arch.

V. The Excretory Organs.
1. Note the shape and position of the kidneys.
2. Trace the ureter from kidney to bladder.
3. Find the adrenal bodies.
4. Take out the kidneys. Cut one kidney into halves lengthwise and horizontally, and the other into halves transversely (corresponding to these planes as the organs lay in the body).

Note:
  (a) The cortex.
  (b) The medulla.
  (c) The pelvic cavity.

Diagrams!

VI. The Reproductive Organs.
1. Identify ovaries or testis (position best found by following out the spermatic artery or vein).
   (a) Ovary: note the Fallopian tube and funnel, the uterus, and the vagina.
   (b) Testis: note its relation to the scrotal sac. Carefully slit open one scrotal sac. Look for the **epididymis**. Trace the vas deferens.

With bone forceps or knife cut through the **symphysis pubis.** *In the female open the vagina just below the uteri by a lateral cut, and continue this to the external opening.*

(c) Find the urethra and open it to the bladder.
(d) Note the orifices of the ureters.
(e) Examine the openings of the uteri and slit one open.

In the male open the urethra by a cut beginning at the external opening and continue the cut to the bladder.

(f) Observe the **uterus masculinus**.
(g) Note the relation of the vasa deferentia to the ureters.

Make a diagram of a view from the side showing the relations of ureter, bladder, vas deferens (oviduct), urethra (vagina).

VII. Find the veins and arteries of the posterior abdominal region and trace those of one side into the leg.

VIII. Review the lobes and vessels of the liver and take it out.

IX. The Nervous System.
   A. The Sympathetic Nervous System.
      A chain of ganglia extending from head to tail close to the vertebral column. **Rami communicantes** connect them with the spinal nerves. The ganglia of one side are also connected by commissures with those of the other side.
   B. The Peripheral Nervous System, nerves passing out from the brain and from

the spinal cord. Note especially at the level of the shoulder:
1. The brachial plexus, made up of the 5th to the 8th cervical and the 1st thoracic nerves inclusive. It gives off four chief nerves; follow them out and see what regions they innervate.

In the region of the leg:
2. The lumbro-sacral plexus, made up of the 5th to the 7th lumbar and of the 1st to the 3d sacral nerves, inclusive. It gives off three chief nerves; follow them out.

C. The Brain.
1. Externals.

With bone forceps cut away as much of the skull as may be necessary to free the brain. Take it out and put it into a dish of formic aldehyde solution or of alcohol.

Find the parts visible on the surface.
On the dorsal side:
(a) The olfactory lobes.
(b) The cerebral hemispheres.
(c) The pineal gland.
(d) The cerebellum.
(e) The medulla, etc.
On the ventral side:
(a) The olfactory lobes.
(b) The cerebral hemispheres.
(c) The optic nerves, chiasma, and tracts.

(d) The pituitary body.
(e) The corpus mammillare.
(f) The crura cerebri.
(g) The pons Varolii.
(h) The cerebellum.
(i) The ventral pyramids.
(k) The medulla.

Sketch the dorsal and the ventral aspect.

2. Section.
Remove the right half of the cerebellum by a median longitudinal cut.
   (a) See the relation of the parts beneath (4th ventricle, corpora quadrigemina) without further cutting.

Remove the other half of the cerebellum.
   (b) The cerebrum.

Make a horizontal cut in the right hemisphere to expose the lateral ventricle.

Observe the hippocampus and see the succession of the optic thalami, the corpora quadrigemina, the third and fourth ventricles.

   (c) The cerebrum, left hemisphere.

Make a frontal section half-way from front to back, or cut off a series of thin sections until the same region is reached, and observe again the same parts.

X. The eye.
Divide one into two lateral halves and the other into anterior and posterior halves.

Find:
1. The sclerotic coat.
2. The choroid coat.
3. The retina.
4. The ora serrata.
5. The ciliary processes.
6. The blind spot.
7. The tapetum.
8. The lens.
9. The iris.
10. The pupil.
11. The humors.

XI. 1. The Ear. Identify as much as possible.
2. The Teeth. Examine their form, position, and roots ~~etc.~~ after pulling them out.

~~XII. Joints.~~
~~1. Hip and knee. Cut into them and see their construction.~~
~~2. Remove the muscle, etc., from two cervical vertebræ and examine their articulation.~~

EMBRYOLOGY.
FROG.

I. **Two-, four-, and eight-celled stages.**[*]
*Study in a watch-glass with a hand-lens. Do not touch the eggs; if it is necessary to turn them over, do so with a stream from a pipette or roll them over by tipping the watch-glass.
Make three sketches of each stage:*
1. *From the dark pole.*
2. *From the light pole.*
3. *From the side half-way between 1 and 2.*

II. **Gastrula**, surface view. Three stages.
Examine as in I.
Note the formation of the **blastopore**.
Sketch.

III. **Blastula.** Prepared section. *Sketch.*
Study with microscope.

IV. **Gastrula.** Prepared section. *Sketch.*
Study with microscope.

V. Closure of the blastopore and the formation and closure of the **medullary folds.** Two stages. *Sketch.*
Study in a watch-glass with a hand-lens.

VI. The appearance of the **adhesive glands**, the eyes, the nasal pits, the mouth, and the gills; and the formation of the **operculum.** Four stages. *Sketch.*

---

[*] When preserved in a dilute aqueous solution of formic aldehyde (1–2 per cent., see page 95) the egg-membranes retain their transparency and the egg has a perfectly normal appearance.

Study in a watch-glass with a hand-lens.

VII. Observe the circulation in the gills of a living tadpole.

Examine in a little water in a watch-glass with microscope, AA 2.

VIII. The early embryo. Prepared transverse section. *Study with AA 2 and D 2.*
 1. The ectoblast (two-layered).
 2. The **neural tube.**
 3. The **notochord.**
 4. The **archenteron.**
 5. The mesoblast.
    (*a*) **Somatic.**
    (*b*) **Splanchnic.**

Sketch, much enlarged.

IX. The dissection of the tadpole.

Dissect under water. Pin the tadpole, dorsal side down, through the base of the tail and the roof of the mouth.

Observe:
 1. The mouth and the horny teeth.
 2. The **eyes.** Are eyelids present?
 3. The **water-pore** from the branchial chamber. On which side is the pore?
 4. The anus.

Open the abdomen by a median incision of the ventral body-wall.

Note:
 5. The coiled intestine. Uncoil it.

EMBRYOLOGY.

6. The liver.
7. The pancreas.
8. The lungs.
9. The heart.
10. The kidneys, **pronephros** or **mesonephros**.
11. The reproductive organs.

Push a bristle into the water-pore and then carefully open the branchial chamber.

12. Examine the internal gills and arches.
13. Count the number of arches.
14. Look for the first pair of feet.
15. Open the pericardium.
16. Find:
    (a) The ventricle.
    (b) The auricles.
    (c) The sinus venosus and its branches.
    (d) The aortic bulb, divided into two parts and each of these again into three parts. Trace these to the gill-arches. Note that the branch to the third arch divides again.

Open the mouth and the pharynx by a ventral longitudinal cut. See:

17. The sieve-like arrangement of the gill-arches.
18. The dorsal aorta, double in the gill region, one on each side receiving the vessels from the gills.
19. Find the posterior feet.
20. Of what structures is the tail made up?

CHICK.
I. The Egg.
  1. Raw egg.
  Holding the egg in the left hand with its rounded end toward the right, make a small opening (by tapping with the points of the large scissors) in the rounded end.
  Note:
  (a) The shell.
  (b) Two membranes.
  (c) The air-space between the two membranes.
  With large scissors cut an oval piece from the top of the shell.
  Look in and identify the parts:
  (d) The white, with **chalazæ**.
  (e) The yolk with the **blastoderm**.
  Diagram !
  Enlarge the opening and gently pour the egg into a dish.
  Does the yolk mingle with the white?
  2. Boiled egg.
  Remove the shell, finding again the membranes and the air-space.
  With the rounded end to the left try to take off the white by unwinding it spirally from left to right.
  Find the blastoderm. With a razor divide the yolk into halves, cutting through the middle of the blastoderm. Try to see:

(*f*) The urn in the centre,—its neck extending to the blastoderm.

(*g*) The alternate layers of white and of yellow yolk.

Diagram !
In the following work the students should leave the slides in the trays on their tables at the end of the laboratory period. If a particular slide is wanted again, its number should be entered by the student on the card that is hanging up for the purpose.

II. The Unincubated Blastoderm.
Examine first with AA 2, *then with D* 2.
1. Surface view. Observe the relative size of the cells in the middle and at the periphery.
2. Section.
Sketch with D 2 *the cell-outlines in a small portion of* (1) *and* (2) *accurately, and on a large scale.*

III. The Early First Day.
1. The **primitive streak.** Surface view.
Study with AA 2. *Sketch carefully.*
2. The medullary folds. Surface view.
Study with AA 2, *then with D* 2. *Sketch carefully.*
3. Sections :
(*a*) Primitive streak. (Single section.)
(*b*) Medullary folds. (Single section.)
(*c*) The early first day chick. (Complete series.)

Study (*a*) and (*b*); then begin at the *posterior* end (why?) of (*c*), and going toward the anterior find the regions of the primitive streak and medullary folds. Observe in the medullary-fold region the notochord and the mesoblast. In which region are the medullary folds most nearly closed?

After studying (*c*), sketch (*a*) and (*b*) with AA 2, *after* having examined them with D 2. Sketch the outlines of the different germ-layers with a black pencil and fill in the space between with *faint* color, using always the same color for the same germ-layer.

IV. The Late First Day.
1. Surface. *Sketch with AA* 2.
2. Sections.
Begin at the posterior end of the series.
Observe:
(*a*) The primitive streak.
(*b*) The medullary folds.
(*c*) The mesoblast.
  Somatic layer.
  Splanchnic layer.
  Mesoblastic somites.
(*d*) The notochord.
(*e*) The closure of the medullary folds to form the neural canal.
(*f*) The closure of the archenteron.

(g) The formation of the head by lateral and anterior folds.

Sketch five or more different regions.

V. The Second Day.
1. Surface.
   Observe:
   (a) The **area vasculosa**.
   (b) The **sinus terminalis**.
   (c) The amnion.
   (d) The vesicles of the brain.
   (e) The eye.
   (f) The ear.
   (g) The lateral outline of the embryo.
   (h) The **omphalo-mesenteric** veins and the heart.
   (i) The dorsal aortæ (?).
   After studying the heart and veins from the dorsal side, turn the slide over and examine the ventral aspect.

Sketch under AA 2.

2. Sections.
   Beginning at the posterior end observe the structures seen in the first day (see IV. 2. (a–g.), and in addition:
   (a) The **Wolffian ducts**.
   (b) The dorsal aortæ.
   (c) The spinal ganglia.
   (d) The omphalo-mesenteric veins.
   (e) The heart.
   (f) The **auditory vesicles.**

(*g*) The aortic arches. What is their relation to the pharynx?
(*h*) The **optic vesicles**.
(*i*) The amnion.
Sketch six different regions.

VI. The Third Day.
1. Surface.
Note the curvature of the embryo; upon which side does it lie?
Note the **cranial flexure**.
Review the structures seen in the second day and observe their further development.
New structures:
(*a*) The anterior vitelline vein. What is the course of the blood within it?
(*b*) The gill-slits.
(*c*) The lens of the eye.
(*d*) The nasal pits.
(*e*) The **allantois**. (Possibly not yet visible.)
Sketch.
2. Sections.
Begin in the middle region (where the archenteron is still open). Make one sketch. Then work backward into the tail-fold and make out its structure *completely*.
Find the section where the dorsal aortæ pass out into the area vasculosa.
What is the origin of the allantois?
At least two sketches desirable.

EMBRYOLOGY. 151

Work thence forward through the head-fold and make out all its relations.

Observe: the vitelline veins, the ducts of Cuvier, the heart (follow out its turns), the pericardial cavity, the aortic arches.

Now begin at the front end of the series and, remembering the cranial flexure, make out the various parts of the brain, the development of the eye, the internal ear, the olfactory pits, the gill-slits (in the neck region), the archenteron and its derivatives: the lungs, the liver, and the pancreas (?).

Study also the origin of the cranial nerves and the Wolffian ducts and their tubules.

Sketches to show these structures.

3. The living embryo.

With the points of strong scissors pierce the broad end of the shell to allow the air to escape. Place the egg in warm normal salt solution (38 degrees C.,) cut away or break away with forceps the upper part of the shell, and examine in situ.

Observe with a hand-lens:

(a) The blastoderm and its areas.
(b) The position and form of the embryo.
(c) The heart, etc.

With fine sharp scissors cut through the blastoderm entirely around the outside of the area vasculosa, float it off into a thin watch-glass, care-

fully remove from the salt solution, and examine with a hand-lens and with A A 2.

Study at first the circulation, the gill-slits, and the movements of the heart. Then review as far as possible the general anatomy.

VII. The Fourth Day.
 1. The formation of the face.

With a razor cut off the head close behind the last gill-slit. With fine forceps remove the parts of the heart, etc., that are attached to the gill-region. Fix the head, dorsal side down, in a small dish of alcohol, and with a hand-lens study and sketch the ventral and lateral aspects.

Find:
 (a) The **naso-frontal** process.
 (b) The nasal pits.
 (c) The eyes.
 (d) The mouth.
 (e) The **superior maxillary process**.
 (f) The visceral arches.

What structures are visible in section on the cut surface.

VIII. Older Stages.
 Examine for the amnion and the allantois, and for the development of the face, the limbs, and the feathers, and for the changes of the **yolk-stalk**.

INDEX.

Abdomen, 67, 128
Abdominal ganglia, 73
Adductor muscles, 58, 59
Adductor muscle, anterior, 59
Adductor muscle, posterior, 59, 61, 65
Adhesive glands, 143
Adrenal bodies, 106, 123, 138
Aerial hyphæ, 40
Air-bladder, 114
Air-bubbles, 6
Air-sacs, 117, 118
Air-space, 146
Albumen, 146
Aleurone, 90, 91
Alimentary canal, 16, 17, 23
Allantois, 150, 152
Amaryllidaceæ, 86
Amaryllideæ, 86
Amaryllis, 86
Amnion, 149, 150, 152
Ampelopsis 80, 81
Anal spot, 30, 33
Angiosperm, 80
Annual rings, 79
Annular vessels, 81, 82
Annulus, 11, 43
Antennæ, 67, 74
Antennules, 67, 74
Anterior adductor muscle, 59
Anterior commissure, 125
Anterior folds, 149
Anterior median seminal vesicle, 19
Anterior nares, 96

Anterior retractor muscle, 59
Antero-posterior differentiation, 14
Anther, 85
Antheridium, 12, 77
Antibrachium, 96, 129
Anus, 15, 61, 67, 113, 128, 144
Aortic arches, 16, 103, 104, 150, 151
Aortic bulb, 115, 145
Aperture, exhalent, 59
Aperture, inhalent, 59
Appendages, 67, 73, 74
Appendages, pyloric, 114
Aqueous humor, 126, 142
Archegonium, 12, 77
Archenteron, 144, 148, 151
Arches, aortic, 16, 103, 104, 150, 151
Arches, visceral, 152
Area vasculosa, 149, 150, 151
Arteries :
    abdominal superior, 69
    antennary, 69
    aorta, 122, 134
    dorsal aorta, 104, 115, 122, 123, 134, 136, 145, 149, 150
    ventral aorta, 115
    aortic arches, 16, 103, 104, 150, 151
    brachial, 122, 135
    carotid, 104
    carotid arch, 104
    caudal, 123
    cephalic circle, 115

INDEX.

Arteries, cœliac, 104, 123, 135
  cœliaco-mesenteric, 104
  common carotid, 122, 135
  cutaneous, 104
  femoral, 122, 136
  genital, 105
  hepatic, 69
  hypogastric, 105
  iliac, 136
  iliac common, 105
  iliac external, 136
  iliac internal, 123, 136
  ilio-lumbar, 136
  innominate, 122, 134
  internal mammary, 135
  lingual, 104
  mesenteric, 105
  mesenteric anterior, 122, 135
  mesenteric inferior, 105
  mesenteric posterior, 123, 136
  ophthalmic, 69
  ovarian, 135
  pectoral, 122
  pulmonary, 123, 136
  pulmonary arch, 104
  renal, 105, 122, 135
  sacral, 136
  sciatic, 105, 122
  spermatic, 135
  sternal, 69
  subclavian, 104, 122, 135
  thoracic, 135
  vertebral, 104, 135
Ascospores, 37,
Asexual generation, 7, 77
Auditory openings, 67
Auditory vesicles, 149
Auricle, 61, 65, 100, 115, 120, 131, 145
Auriculo-ventricular valves, 137
Axial filament, 34
Axial strand, 76
Axillary air-sacs, 118

Bacteria-cultures, 46, 47
Barbicels, 127
Barbs, 54, 127

Barbules, 127
Basidia, 40
Bast-cells, 81
Bast-fibres, 9
Bast-vessels, 9
Bean, 87, 90, 91, 92
Bilateral symmetry, 14·
Bile-duct, 100, 130
Bladder, 99, 139
Blastoderm, 146, 147, 151
Blastopore, 143
Blastula, 143
Blind spot, 126, 142
Blood-vessel, dorsal, 13, 16
Body, 128
Body-cavity, 16
Body-wall, 16, 98
Bojanus, organ of, 62, 65
Brachial nerve, 110
Brachial plexus, 110, 124, 140
Brachium, 96, 129
Brain, 107, 124, 140, 141
Brain-cavities, 109
Brain, vesicles of, 149, 151
Branchiæ, 60, 64, 65
Branchial chamber, 60, 65, 144
Branchial nerve, 62
Branchio-cardiac groove, 66
Branchiostegites, 66
Brownian movement, 45
Bubbles of air, 6
Buccal cavity, 97
Buccal pouch, 17, 21
Buckwheat, 87, 88, 92
Bud, 36, 53, 78, 82
Bud-scales, 82, 83
Bulbus, 115, 145
Bundle, fibrovascular, 8, 9, 10, 79, 80, 81, 82
Bundle-sheath, 9, 82

Calamus, 126
Calciferous glands, 18, 19
Calyptra, 77
Calyx, 84, 85
Cambium, 80
Canal, sternal, 72

INDEX. 155

Capsule, 11, 77
Capsulogenous glands, 14
Carapace, 66
Castor-oil bean, 87, 88, 91
Cauda equina, 108
Caulicle, 89
Cavities of brain, 109
Cavity, buccal, 97
Cavity, pericardial, 61, 62, 115
Cell-division, 29
Central nervous system, 20, 72, 106
Cephalic circle, 115
Cephalothorax, 66
Cerebellum, 108, 124, 140, 141
Cerebral ganglia, 21, 23, 66
Cerebral hemispheres, 107, 108, 109, 124, 125, 140, 141
Cerebral peduncles, 126
Cerebro-splanchnic commissure, 63
Cervical groove, 66
Chalaza, 87
Chalazæ, 146
Chamber, branchial, 60, 65
Chamber, cloacal, 60, 61
Chamber, suprabranchial, 60, 65
Chemical experiments, 32, 35, 37
Chloragogue cells, 18
Chordæ tendineæ, 137
Choroid coat, 111, 126, 142
Choroid plexus, 108
Chromatophore, 27, 29, 49
Cilia, 30, 31, 32, 33, 34, 35
Ciliary processes, 126, 142
Ciliated rosettes, 19
Circle, cephalic, 115
Circular vessels, 17
Circumœsophageal commissure, 21, 23, 72
Clitellum, 14
Cloaca, 123
Cloacal chamber, 60, 61
Cnidoblasts, 53, 55, 56
Cnidocil, 53
Cœlom, 16

Commissure anterior, 125
Commissure cerebro-splanchnic, 63
Commissure circumœsophageal, 21, 23, 72
Commissure optic, 125
Commissure posterior, 125
Conidia, 40
Conidiophores, 40
Conjugation, 35
Connective, 85
Contractile vacuole, 24, 31, 34
Conus, 115
Cork, 80
Cornea, 111
Corolla, 84, 85
Corpora quadrigemina, 141
Corpus mammillare, 141
Corpus striatum, 125
Cortex, 138
Cortical layer, 41, 42
Cortical parenchyma, 80
Cotyledons, 88, 89, 92
Coxopodite, 67
Cranial flexure, 150, 151
Cranial nerves, 151
Crop, 17, 116, 119
Crura cerebri, 109, 141
Crus, 96, 129
Crystalloids, 91
Cultures, of bacteria, 46, 47
Currents in entoplasm, 31, 34
Curvature of embryo, 150
Cuticle, 16, 22, 30, 34, 66
Cuvierian ducts, 115, 151

Daffodil, 86
Dehiscence, 12
Diaphragm, 131
Dicotyledon, 80, 91
Differentiation, antero-posterior, 14
Differentiation, dorso-ventral, 14
Digestive gland, 65, 71
Digits, 96, 129
Disk, 33, 35
Dissepiments, 16

INDEX.

Division, cell, 29
Division, endogenous, 28, 37
Division, nuclear, 94
Dorsal blood-vessel, 13, 16
Dorsal fissure, 108
Dorsal pores, 15
Dorso-ventral differentiation, 14
Drops, oil, 6
Ducts of Cuvier, 115, 151
Ducts, Wolffian, 149, 151
Duodenum, 99, 100, 119, 122, 129
Dura mater, 107

Ear, 128, 142, 149, 151
Early embryo, 144
Ectoblast, 144
Ectoplasm, 24, 30, 34
Egg, 146
Egg-capsule, 22
Egg-membranes, 143, 146
Egg-shell, 146
Elder, 82
Embryo, early, 144
Endogenous division, 28, 37
Endophragmal skeleton, 72
Endophragmal system, 72
Endosperm, 88, 89
Entoplasm, 24, 31, 34
Entoplasm, currents in, 31, 34
Epidermis, 7, 9, 10, 11, 75, 80, 81, 83, 84
Epididymis, 138
Epiphragma, 78
Epistome, 33
Erect hyphæ, 40
Eustachian recesses, 97
Excretory openings, 61, 67, 139
Exhalent aperture, 59
Exoskeleton, 66, 73
Eye, 96, 111, 126, 141, 142, 143, 144, 149, 150, 151, 152
Eyelids, 96

Face, 152
Falciform ligament, 117, 119
Fallopian tube, 138
Fat-bodies, 100

Fat-drops, 36, 40
Feather, 126, 127
Feathers, arrangement of, 116
Feathers, development of, 152
Feet, 145
Femur, 96, 112, 129
Fibrovascular bundle, 8, 9, 10, 11, 80, 81, 84
Ficus elastica, 83
Filament, 85
Filament, axial, 34
Filum terminale, 108
Fission, 35
Fissure, dorsal, 108
Flagella, 29
Folds, anterior, 149
Folds, lateral, 149
Folds, medullary, 143, 147, 148
Food-vacuoles, 24, 31, 34
Foot, 52, 60, 65
Foramen of Monro, 109
Formalin, 95
Formic aldehyde, 95, 96, 143
Fossa ovalis, 137
Fourth ventricle, 125
Frond, 7
Frontal spine, 66
Fundamental parenchyma, 8, 9, 81, 82
Funiculus, 87
Funnel, 19, 138
Furcula, 118

Gall-bladder, 99, 113, 130
Ganglion, cerebral, 21, 23, 63
  parieto-splanchnic, 62, 65
  pedal, 63
  spinal, 149
  subœsophageal, 21
  supraœsophageal, 72
  thoracic, 72
  visceral, 62, 65
Gastric teeth, 71
Gastrocnemius muscle, 111
Gastrula, 143
Gemmation, 37
Generation asexual, 7, 77

INDEX.

Generation sexual, 12, 75
Genital openings, 67
Geranium, 84
Gills, 60, 64, 65, 66, 73, 113
Gills, circulation in, 144, 152
Gills, internal, 145
Gill-arches, 145
Gill-clefts, 113
Gill-slits, 150, 151
Gizzard, 17, 119, 122
Gland-cells, 55
Gland, digestive, 65
Gland, pineal, 107, 140
Glands, adhesive, 143
Glands, calciferous, 18, 19
Glands, capsulogenous, 14
Glands, green, 71
Globoids, 91
Glochidia, 64
Glottis, 97, 99
Gonidia, 42
Gonidial layer, 41
Gray, 86
Green glands, 71
Groove, branchio-cardiac, 66
Groove, cervical, 66
Guard-cells, 10, 11, 83
Gymnosperm, 79

Hay infusion, 47
Head, 96, 116, 128, 149
Head-fold, 149, 151
Heart, 61, 68, 98, 100, 114, 119, 131, 136, 137, 145, 149, 151, 152
Hilum, 87
Hip-joint, 142
Hippocampus, 141
Horse-chestnut, 82
Humor aqueous, 126, 142
Humor vitreous, 111, 126, 142
Hyacinth, 83
Hydra viridis, 53, 55
Hyphæ, 40, 42, 43, 44
Hyphæ aerial, 40
Hyphæ erect, 40
Hypocotyl, 92

Hypodermis, 16
Hypostome, 52

Indian corn, 81, 87, 88, 92
India-rubber tree, 83
Indusium, 7, 11
Infundibulum, 125
Inhalent aperture, 59
Intercellular spaces, 84, 90
Interclavicular air-sac, 118
Internode, 7
Interstitial cells, 55, 56
Intestine, 71, 99, 113, 117, 119, 129, 133, 144
Iris, 126, 142

Keel, 117
Kidneys, 108, 114, 123, 138, 145
Knee-joint, 142

Labial palps, 60, 61, 65
Lateral folds, 149
Lateral ventricle, 125
Lamellæ, 43, 44
Lamina, 7
Layer cortical, 41, 42
Layer gonidial, 41
Layer medullary, 42
Layer muscular, 16
Leguminosæ, 86
Lemna, 82
Lens, 111, 126, 142, 150
Ligament, falciform, 117, 119
Liliaceæ, 86
Lily, 86
Limbs, 96, 128, 152
Lips, 128
Liver, 98, 99, 101, 113, 119, 129, 130, 145, 151
Lobe, 7, 85
Lobes, olfactory, 107, 108, 140
Lobes, optic, 108
Lobule, 7
Lumbar plexus, 124
Lumbricus terrestris, 13
Lumbro-sacral plexus, 140
Lungs, 98, 99, 145, 151

Macronucleus, 31, 32
Macrozoöspores, 28
Mammæ, 128
Mandible, 74
Mantle, 58, 59, 65
Mantle-cavity, 60
Manus, 96, 129
Mass, visceral, 60, 61, 65
Material, 27, 39, 48, 58, 66, 78, 83
Maxilla, 74
Maxillary process, superior, 152
Maxilliped, 73, 74
Mechanical experiments, 25
Medulla, 138
Medulla oblongata, 108, 124, 140, 141
Medullary folds, 143, 147, 148
Medullary layer, 42
Medullary rays, 78, 80
Membrane, tympanic, 96
Mesenteron, 71
Mesentery, 99, 117, 129, 130
Mesoblast, 148
  somatic, 144, 148
  splanchnic, 144, 148
Mesoblastic somites, 148
Mesocarpus, 50
Mesonephros, 145
Mesophyll, 11
Metamerism, 14
Methods, 13, 25, 36, 50, 53, 54, 55, 58, 94, 95, 143
Micrometer, stage, 6
Micronucleus, 31, 32
Micropyle, 87, 88
Microzoöspore, 29
Mitral valve, 137
Mnium, 76
Monocotyledon, 81, 92
Monro, foramen of, 109
Motile forms, 28, 29, 35
Mouth, 14, 17, 30, 33, 52, 61, 67, 97, 128, 143, 144, 152
Movement, Brownian, 45
Müller, 86
Muscle, adductor, 58, 59

Muscle, gastrocnemius, 111
  papillary, 137
  pectoralis major, 118
  pectoralis minor, 118
  sartorius, 111
Muscular layer, 16
Mustard, 82, 92
Mycelium, 39, 40, 43, 44

Narcissus, 86
Nares, anterior, 96
  posterior, 97
Nasal pits, 143, 150, 152
Naso-frontal process, 152
Neck, 116, 128, 151
Nematocyst, 53, 54, 55
Nephridia, 16
Nephridial openings, 15
Nerve, brachial, 110
  branchial, 62
  cranial, 151
  olfactory, 107, 108
  optic, 140
  pallial posterior, 62
  sciatic, 110
  septal, 21
  spinal, 110
  spinal thoracic, 124
Nerve-cord, ventral, 21, 23
Nerve-plexus, brachial, 110, 124, 140
  lumbar, 124
  lumbro-sacral, 140
  sciatic, 110, 124
Nervous system, 20, 106, 123, 139, 140
Nervous system, central, 20, 72, 106
  peripheral, 21, 110, 124, 139
  sympathetic, 109, 123, 139
Nettle-batteries, 53, 54
Neural tube, 144
Nitella, 93
Node, 7
Nostrils, 128
Notochord, 144, 148
Nucellus, 87

INDEX. 159

Nuclear division, 94

Œsophagus, 17, 30, 31, 33, 70, 116, 119
Oil-drops, 6
Olfactory lobes, 107, 108, 140
Olfactory nerves, 107, 108, 124
Olfactory pits, 151
Olfactory tract, 107, 108
Onion, 94
Oöphore, 75
Oösphere, 12
Opening, urino-genital, 128
  excretory, 67, 139
  genital, 67
  nephridial, 15
Operculum, 78, 113, 143
Ophthalmic artery, 69
Optic chiasma, 109, 125, 140
  commissure, 125
  lobes, 108, 124, 125, 126
  nerves, 140
  thalami, 125, 141
  tracts, 140
  ventricle, 126
  vesicle, 150
Ora serrata, 126, 142
Organ of Bojanus, 62, 65
Ostia, 68, 69
Ovary, 20, 70, 86, 105, 123, 138
Oviduct, 15, 20, 70, 105, 123, 139
Ovules, 85

Palisade-cells, 84
Pallial cavity, 60
Palps, labial, 60, 61, 65
Pancreas, 99, 119, 122, 129, 145, 151
Pancreas, duct of, 129, 130
Pansy, 87
Papillary muscles, 137
Paraffine thermometer, 25
Paraphyses, 77
Parenchyma, cortical, 80
  fundamental, 8, 9, 81, 82
  sclerotic, 7
  wood, 81

Parieto-splanchnic ganglion, 62, 65
Parmelia, 41
Pea, 87, 88, 90, 91, 92
Pecten, 126
Pectoralis major muscle, 118
Pectoralis minor muscle, 118
Pectoralis secundus muscle, 118
Pedal ganglia, 63
Pedicel, 84
Peduncles, 126
Pelvic cavity of kidney, 138
Pelvis, 112
Pericardial cavity, 61, 62, 114, 151
Pericardial sinus, 68
Pericardium, 145
Perichætium, 75
Peripheral nervous system, 21, 110
Peristome, 33, 78
Pes, 96, 129
Petals, 85
Pharynx, 17, 97, 145, 150
Phloëm, 80, 81
Phloëm-parenchyma, 10
Phloëm-sheath, 9
Physiological experiment, 38
Pia mater, 107
Pileus, 43
Pineal gland, 107, 140
Pinna, 7, 128
Pinnula, 7
Pistil, 84
Pith, 80
Pits, 79, 80
Pituitary body, 109, 125, 141
Plasmolysis, 49
Pleurococcus, 28, 29
Plumule, 88, 89
Pollen-grains, 85, 86
Polytrichum, 75
Pons Varolii, 141
Pontederia, 82
Pores, dorsal, 15
Posterior adductor muscle, 59, 61, 65

Posterior commissure, 125
Posterior median seminal vesicle, 19
Posterior nares, 97
Posterior pallial nerve, 62
Potato cultures, 46, 47
Pouch, buccal, 17, 21
Prebronchial air-sac, 118
Primitive streak, 147, 148
Primordial utricle, 49
Processes ciliary, 126, 142
Pronephros, 145
Prosenchyma, 82
Prosenchyma sclerotic, 8, 9
Prostomium, 14
Prothallium, 12
Protonema, 78
Proventriculus, 119, 122
Pseudopodia, 24, 25
Pteris aquilina, 7
Pupil, 126, 142
Pyloric appendages, 114
Pylorus, 130
Pyramids, 141
Pyrenoids, 49

Quill, 126

Rachis, 127
Radial symmetry, 53
Radicle, 88, 89
Rami communicantes, 139
Rays medullary, 79
Receptacle, 85
Receptacles seminal, 14, 20
Rectal cœca, 119
Rectum, 65, 119
Rete mirabile, 114
Retina, 111, 126, 142
Rhizoids, 41, 75, 76
Rhizome, 7, 8, 9, 10
Rings annual, 79
Root, 7, 82, 92, 93, 94
Root-cap, 82
Root-hair, 82
Root-sheath, 89
Rosettes, ciliated, 19

Rotation, 29

Sacs scrotal, 128, 138
Sartorius muscle, 111
Scale, 6
Scales, 15
Scaphognathite, 73
Scapus, 126
Sciatic nerve, 110
    plexus, 110, 124
Sclerenchyma, 8
Sclerotic coat, 111, 126, 142
    parenchyma, 7, 9
    prosenchyma, 8, 9
Scrotal sacs, 128, 138
Scutellum, 89
Sedgwick and Wilson, 19, 22, 36
Seed-coats, 87, 89
Seeds, 87
Segmentation stages, frog, 143
Semilunar valves, 137
Seminal receptacles, 14, 20
    vesicles, 16, 19
    vesicle, anterior median, 19
    vesicle, posterior median, 19
Sepals, 85
Septal nerves, 21
Septum, 137
Serial symmetry, 141
Seta, 14, 77
Seta-sac, 22
Sexual generation, 12, 75
Shaft, 127
Sheath, 34
Sieve-tubes, 9, 81
Sinus, pericardial, 68
    terminalis, 149
    venosus, 100, 102, 115, 145
Skeleton endophragmal, 72
Somites, 13
Somites mesoblastic, 148
Sorus, 7, 11
Spaces, intercellular, 84, 90
Spermatozoa, 19
Spermatozoid, 12
Spinal cord, 108, 109
    ganglia, 149

Spinal nerves, 110
nerves, thoracic, 124
Spine, frontal, 66
Spiral vessels, 81, 82
Spleen, 99, 114, 119, 122
Sporangium, 7, 11
Spore, 12, 39, 43, 77, 78
Sporophore, 77
Spot, anal, 30
Stage, micrometer, 6
warm, 25, 26
Stalk, 34, 35
Stamens, 84, 85
Starch, 9, 10, 37, 89, 90, 93
Stems, 126
Sterigmata, 40, 44
Sternal artery, 69
canal, 72
Sternum, 117
Stigma, 86
Stipe, 7, 43
Stoma, 10, 11, 83, 84
Stomach, 64, 70, 99, 113, 129
Stomach intestine, 17, 18
Strand, axial, 76
Streak, primitive, 147, 148
Subepidermis, 7, 8, 9, 76, 80, 81
Subœsophageal ganglion, 21
Superior abdominal artery, 69
Supporting lamella, 56
Suprabranchial chamber, 60, 65
Supraœsophageal ganglia, 72
Symmetry, bilateral, 14
radial, 53
serial, 14
Sympathetic ganglia, 110
nervous system, 109
Symphysis pubis, 138
System, endophragmal, 72
nervous, 20

Tadpole, 144
Tail, 116, 128, 145
Tail-fold, 150
Tapetum, 142
Teeth, 97, 142, 144

Teeth, gastric, 71
of peristome, 78
Telson, 73
Testis, 20, 70, 94, 106, 123, 138
Testis, cell-division in, 94
Thalamencephalon, 107
Thallus, 41, 42
Theca, 85
Thermometer, paraffine, 25
Third ventricle, 125
Thoracic ganglia, 72
Thorax, 128, 131
Thymus, 131
Tongue, 97
Trachea, 116, 131
Tracheæ, 10, 81
Tracheids, 10, 79, 80, 81
Tradescantia, 93
Trichocyst, 32
Tricuspid valve, 137
Truncus arteriosus, 100, 103
Trunk, 96, 116, 128
Tuber cinereum, 109
Tympanum, 96, 97
Tympanic membrane, 96
Typhlosole, 18, 64

Umbilicus inferior, 127
superior, 127
Ureter, 106, 123, 138, 139
Uterus, 138
Uterus masculinus, 139
Urethra, 139
Urino-genital opening, 128
pouch, 123
Urn, 147
Utricle, primordial, 49

Vacuole, 40
Vacuole, contractile, 24, 31, 34
food, 24, 31, 34
water, 24, 31, 34
Vagina, 138
Valve, mitral, 137
semilunar, 137
tricuspid, 137
Vane, 127

Vas deferens, 15, 19, 70, 123, 138, 139
Vasa efferentia, 106
Vein, abdominal anterior, 98, 100, 101, 102
  azygos, 132
  brachial, 103, 120
  caudal, 120
  coccygeo-mesenteric, 120, 121
  coronary, 133
  epigastric, 119
  femoral, 101, 120
  gastric, 102, 121
  gastro-duodenal, 121
  genital, 102
  hepatic, 102, 113, 114, 121, 133
  hepatic portal, 102
  hepatic portal system, 121
  hypogastric, 120, 121
  iliac, 120, 134
  iliac, external, 134
  iliac, internal, 120, 134
  ilio-lumbar, 133
  innominate, 103
  intercostal, 132
  intestinal, 102
  jugular, 102, 103, 120, 133
  lingual, 103
  mammary internal, 132
  maxillary inferior, 103
  mesenteric anterior, 121
  mesenteric posterior, 120, 121
  musculo-cutaneous, 98, 103
  omphalo-mesenteric, 149
  ovarian, 133
  pectoral, 120
  pelvic, 101
  portal, 121, 134
  precaval, 115
  pulmonary, 122, 134
  renal portal, 101
  renal, 102, 121, 133
  sciatic, 101, 121
  spermatic, 133
  splenic, 102
  subclavian, 103, 133

Vein, subscapular, 103
  vena cava, 62, 65
  vena cava anterior, 120, 132, 133
  inferior, 102
  posterior, 120, 133
  superior, 102, 103
  vertebral, 120, 132
  vitelline, 150, 151
  vitelline, anterior, 150
Velum, 43
Ventral nerve-cord, 21, 33
Ventral pyramids, 141
Ventricle, 61, 65, 100, 115, 120, 131, 145
Ventricle, fourth, 108, 109, 125, 141
  lateral, 109, 125
  optic, 109, 126
  third, 109, 125, 141
Vertebra, 142
Vesicles, auditory, 149
  of brain, 149, 151
  optic, 150
  seminal, 16, 19
Vessels, annular, 81, 82
  circular, 17
  spiral, 81, 82
Vestibule, 30, 33
Vexillum, 127
Vibrissæ, 128
Violaceæ, 86
Violarieæ, 86
Violet, 86, 87
Visceral arches, 152
  ganglion, 62, 65
  mass, 60, 61, 65
Vitellus, 146, 147
Vitreous humor, 111, 126, 142

Warm stage, 25, 26
Water-pore, 144
Water-vacuole, 24, 31, 34
Wheat, 87, 89, 90, 92
Whiskers, 128
Whorl, 84
Wickersheimer's fluid, 95

INDEX.

Wolffian ducts, 149, 151
Wood-parenchyma, 10, 81

Xylem, 10, 80, 81

Yolk, 146, 147

Yolk-stalk, 152
Zoöchlorella, 55
Zoöglæa, 46
Zoöspore, 28, 29
Zygnema, 50

www.ingramcontent.com/pod-product-compliance
Lightning Source LLC
Chambersburg PA
CBHW020306170426
43202CB00008B/514